Process Engineering: Facts, Fiction, and Fables

Scrivener Publishing
100 Cummings Center, Suite 541J
Beverly, MA 01915-6106

Publishers at Scrivener
Martin Scrivener (martin@scrivenerpublishing.com)
Phillip Carmical (pcarmical@scrivenerpublishing.com)

Process Engineering: Facts, Fiction, and Fables

Norman P. Lieberman

Scrivener
Publishing

WILEY

This edition first published 2017 by John Wiley & Sons, Inc., 111 River Street, Hoboken, NJ 07030, USA and Scrivener Publishing LLC, 100 Cummings Center, Suite 541J, Beverly, MA 01915, USA
© 2017 Scrivener Publishing LLC
For more information about Scrivener publications please visit www.scrivenerpublishing.com.

Wiley Global Headquarters
111 River Street, Hoboken, NJ 07030, USA

For details of our global editorial offices, customer services, and more information about Wiley products visit us at www.wiley.com.

Limit of Liability/Disclaimer of Warranty
While the publisher and authors have used their best efforts in preparing this work, they make no representations or warranties with respect to the accuracy or completeness of the contents of this work and specifically disclaim all warranties, including without limitation any implied warranties of merchantability or fitness for a particular purpose. No warranty may be created or extended by sales representatives, written sales materials, or promotional statements for this work. The fact that an organization, website, or product is referred to in this work as a citation and/or potential source of further information does not mean that the publisher and authors endorse the information or services the organization, website, or product may provide or recommendations it may make. This work is sold with the understanding that the publisher is not engaged in rendering professional services. The advice and strategies contained herein may not be suitable for your situation. You should consult with a specialist where appropriate. Neither the publisher nor authors shall be liable for any loss of profit or any other commercial damages, including but not limited to special, incidental, consequential, or other damages. Further, readers should be aware that websites listed in this work may have changed or disappeared between when this work was written and when it is read.

Library of Congress Cataloging-in-Publication Data
ISBN 978-1-119-37027-7

Cover image: Norman P. Lieberman
Cover design by Kris Hackerott and Roy Williams

Set in size of 11pt and Minion Pro and 12pt Comic Sans MS by Exeter Premedia Services Private Ltd., Chennai, India

10 9 8 7 6 5 4 3 2 1

To Allen and Irene Hebert, whose dedication and determination have been tirelessly applied to assemble this text. And who jointly originated the concept for assembling my cast of cartoon characters into book format.

Contents

Other Books By Norman Lieberman

   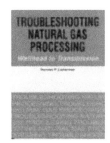

REVAMPS & TROUBLE SHOOTING

Process Chemicals Incorporated

5000A West Esplanade PMB 267
Metairie, LA 70006
PH: (504) 887-7714 FAX: (504) 456-1835
Email: norm@lieberman-eng.com
www.lieberman-eng.com

Introduction

I started work as a process engineer for the American Oil Company in 1965. Now, after 52 years, I'm still a process engineer. Still working in the same way, on the same problems:

- Distillation Tray Efficiency
- Shell & Tube Heat Exchangers
- Thermosyphon Reboilers
- Draft in Fired Heaters
- Steam Turbine Operation
- Vacuum Steam Ejectors
- Centrifugal Pump Seals
- Surge in Centrifugal Compressors
- Reciprocating Compressor Failures
- Process Safety
- Fluid Flow

Most of what I need to know to do my job, I have still to learn. And I'm running out of time! So, with the help of my little friends in this book, I've recorded what I have learned so far. I hope this will help you in solving process problems.

The difficulty of being a process engineer is that our job is to solve problems. Not with people, but with equipment. Within minutes, or hours, or days, the validity of our efforts are apparent. More like plumbing, less like other branches of technology.

Most things I've tried as a process engineer haven't worked. But those that have been successful I remember, and use again. And it's insights from these successful plant trials and projects that I have shared with you in my book.

One thing's for certain. The money paid for this book is nonrefundable. But should you have process questions, I'll try to help.

Norm Lieberman
1-504-887-7714
norm@lieberman-eng.com

PART I

CHAPTER ONE

PROCESS OPERATIONS & DESIGN

CARL & CLARE

Hello! I'm Clare! I work for Carl. We troubleshoot refinery process equipment! We're a team!

Hi! I'm Carl! I know everything, because I'm really, extremely, smart! Clare is my associate!

INCREASING COOLING WATER FLOW THRU AN ELEVATED CONDENSER OR COOLER

 Clare! Let's open the cooling water outlet valve to get more water flow.

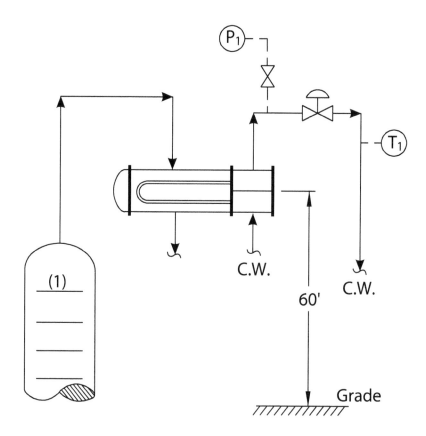

No, Carl! The Condenser is 60 feet above grade. The pressure at P_1, is under vacuum! Opening that valve will give us less cooling water flow!

NO! Opening a valve will always increase flow!

Sorry, Carl! Opening that valve reduces the pressure at P_1, further below the atmospheric pressure. This causes the air to flash-out of the cooling water, which chokes back water flow!

 Clare! WRONG! I'm really smart! Anyway, where's the test to prove you're right?

OK. I'll close the valve and you'll see the temperature at T_1 will go down. But don't close it too much! Otherwise, you will throttle the water flow. Then, T_1 will get hotter!

 But Clare! How do I know how to adjust that stupid valve?

Carl, dear! Set the valve to hold a back-pressure of about 3" Hg. That's minus 0.10 atmosphere. At 100 °F, that will stop air evolution from the water, but not throttle the water flow too much!

HOT VAPOR BY-PASS PRESSURE CONTROL

 Clare! Close the hot vapor by-pass valve! We need to lower the tower pressure. Do it now!

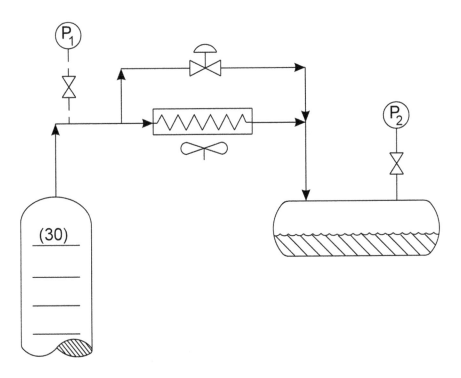

 Sorry, Carl! When I closed the valve the tower pressure went up ... not down!

That woman is a fool! Why did I ever hire her?

No, Clare! Closing the valve will cool off the reflux drum! The pressure at P2 will drop, and draw down the pressure at P1. Understand?

But Carl! How about the pressure drop across the air cooler? It increases as more flow is forced through it. True, the pressure at P2 will always fall! But the pressure at P1 may go up or down—depending on the air cooler DP!

But, but...? Closing the hot vapor by-pass is supposed to lower the tower pressure, according to my design manual!

But suppose the tubes get full of salts and scale? Then what? Also, Carl, we now have a vacuum in the reflux drum, which can be quite dangerous! Air could be sucked into the drum and an explosive mixture could form! Don't forget there's pyrophoric iron sulfide deposits ($Fe(HS)_2$) in the drum! They'll auto-ignite at ambient temperatures!

I really need to get a new assistant! That woman is so annoying! She never shows proper respect! Maybe I could hire a real smart control engineer from MIT?

STALLING A THERMOSYPHON REBOILER

 Clare! Open the steam supply valve! Quick! We need more reboiler heat. The reflux drum is going empty!

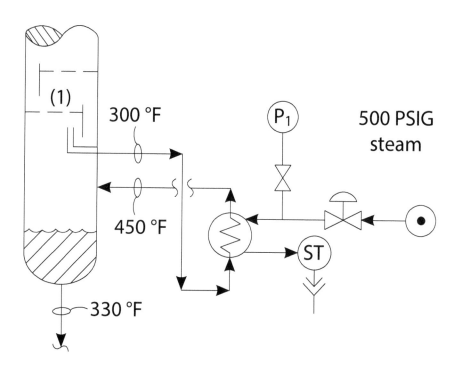

Sorry, Boss! That won't help! The Once-Thru Thermosyphon Reboiler is STALLED OUT!

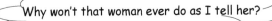 Why won't that woman ever do as I tell her?

 Clare! More steam flow will have to give us more heat to the reboiler! Open that valve!

Opening the steam valve will not increase steam flow when the reboiler is STALLED-OUT!

 STALLED-OUT? What does that mean?

Stalled-out means heat duty is limited by the process flow to the tube-side of the reboiler! The process flow rate to the reboiler is real low now and limiting the steam condensation rate!

 How do you know that, Clare? Do you have X-ray vision?

Carl! Look at the reboiler outlet. It's 450°F! The tower bottoms are only 330°F. Most of the 300°F liquid from tray #1 is leaking past the draw pan, and dumping into the bottoms' product!

OK, Clare, OK! But still, the steam inlet valve is only 50% open! Won't opening it 100% help some?

No, Sir! The pressure at P1 on the steam inlet line is 500 PSIG! The same as the steam supply pressure. There is zero DP across the steam supply valve. The valve position, with no DP, is IRRELEVANT!

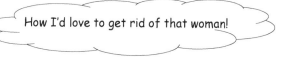

How I'd love to get rid of that woman!

I guess we should have used a total trap-out chimney tray for tray #1! I remember you suggested that last year, Clare. Perhaps you'd like a transfer to the Process Design Division? They would probably love to have you! I remember that in the old days we had bubble cap trays, which could never leak and cause this loss in thermosyphon circulation, or stalling-out.

OPTIMIZING FRACTIONATOR PRESSURE

Clare! The best way to optimize tower pressure is to target for the lowest pressure!

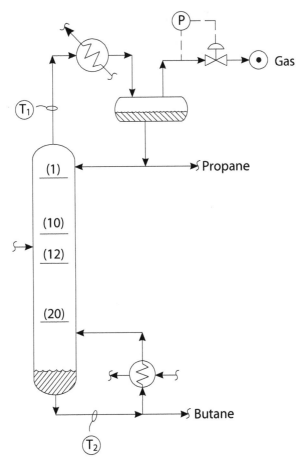

Why is that, Carl?

Because, Clare, as we learned at university, the lower the pressure, the greater the RELATIVE VOLITILITY between propane and butane!

But Carl! Suppose the lower tower pressure causes entrainment? Then, a lower pressure will reduce tray separation efficiency and make fractionation worse!

That woman is always so negative!

Well! What do you suggest? It takes too long to wait for lab sample results.

Carl, I suggest:

1. At a constant reflux rate, start lowering the fractionator pressure.
2. Now, watch the delta T (T1 - T2).
3. That tower pressure, that maximizes delta T, will give the best split between butane and propane. But make the moves slowly!

I think Clare's been spending too much time hanging out with the plant operators!

She's starting to think like them!

Clare! You really should take a more positive attitude towards your engineering degree, and show more respect for the principle of relative volatility!

ADJUSTING STEAM TURBINE SPEED TO MINIMIZE STEAM CONSUMPTION

 Clare! Always run steam turbines at a constant speed! In the U.S., 3,600 rpm, Europe 3,000 rpm!

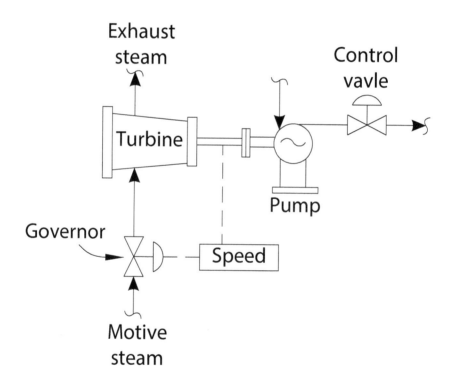

I'm sorry, Carl! But I don't agree!

Exactly what's your problem? All process plants run their turbines at 5% below their maximum rated speed. That's best!

Carl! The best way to set the speed of the turbine is to slow the turbine down until the control valve on the discharge of the pump it's driving is in a mostly wide-open but still controllable position!

All to what purpose?

Well, Carl, for each 3% speed reduction, the steam required to drive the turbine will fall by 10%. Work varies with speed cubed: $W \sim (speed)^3$...That's the Affinity Law!

That woman! Now she's making up her own "LAWS"!

Thanks, Clare! I'll write new instructions for the operators!

Don't tell Carl, but during the 1980 strike in Texas City, Norm saved 30% of the driver steam at the sulfur plant by adjusting steam turbines! All in a single twelve-hour shift!

Carl! Actually, you can get rid of the control valve on the discharge of the pump, and run just on governor speed control, to adjust the upstream level or the pump discharge flow and pressure. That will save even more of the motive steam.

STEAM CONDENSATE DRAINAGE FROM REBOILERS BLOWING CONDENSATE SEAL

Clare, listen up! Open that condensate drain valve more. We need more heat to the reboiler!

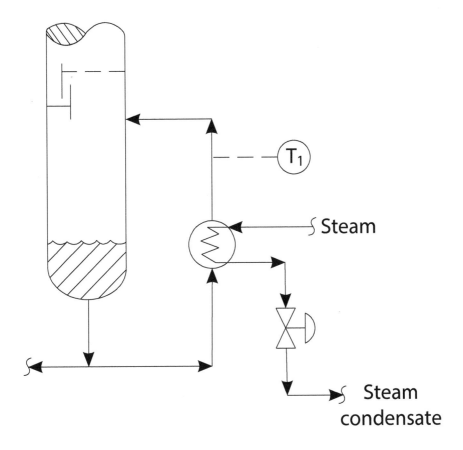

Steam

Steam condensate

Sorry, boss! I think opening the condensate drain valve more will reduce reboiler duty! I'm not too sure!

Not sure? Can't you just follow my instructions for once? It's getting late!

Carl, here's the problem! If I open the drain valve too much, we'll blow the condensate seal. Steam will blow through the reboiler tubes, without condensing. If I open the valve too little, we will suffer from condensate back-up!

I'm confused! Then how do we know whether to open or close that stupid condensate drain valve?

Well, if we close the valve, and T1 goes up, it means we were previously blowing the condensate seal! If we close the valve and T1 goes down, it means we were previously suffering from steam condensate back-up!

I wonder if I can get Clare to buy me lunch today?

OK! Let's optimize the drain valve position to maximize the reboiler outlet temperature, and go to lunch. I'm hungry! Seems like it's all a balance between condensate back-up and blowing the condensate seal. Kinda like eating too much or too little.

EFFECT OF REFLUX ON FRACTIONATOR TOP TEMPERATURE

 Clare, dear! To lower the tower top temperature, one should always raise the reflux rate. It's a basic idea of Process Control!

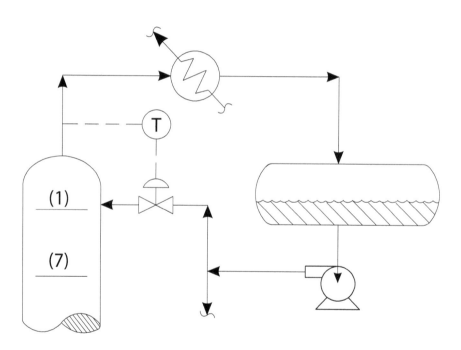

Sorry to disagree again, Carl! It ain't necessarily true!

Now Clare! Read any distillation textbook.
They all agree with me! What's your problem?

It's this: If the top tray is at its flood point,
raising top reflux must increase reboiler duty,
because the reflux comes from the reboiler.
This will increase vapor flow to the top tray!

So what? The reflux will cool off that
extra vapor! The reflux will knock back the
heavier components in the vapor. What's your
problem, Clare?

What you say is true, boss. Up to a point.
The INCIPIENT FLOOD POINT! Above that
point, extra vapor promotes ENTRAINMENT.
The result is droplets of heavy liquid blow
up through the trays, increasing the boiling
range of the top product and temperature at
the top of the tower.

That's BAD! Because then the tower top
temperature will go up, as the reflux rate is
increased. Worse, the top reflux rate, and
reboiler duty will then increase more, and
make the problem even worse!

Yes, Carl! I call this getting caught up in a POSITIVE FEED-BACK LOOP! The reflux needs to be switched to manual and partly closed to break the feed-back loop.

That woman has no grasp of basic process control strategy! How did she ever get an engineering degree? She's always getting confused by improper and prohibited equipment malfunctions!

CENTRIFUGAL PUMP HEAD VS. FLOW
PERFORMANCE CURVES

Clare! Where did you get that pump curve from? It's wrong!

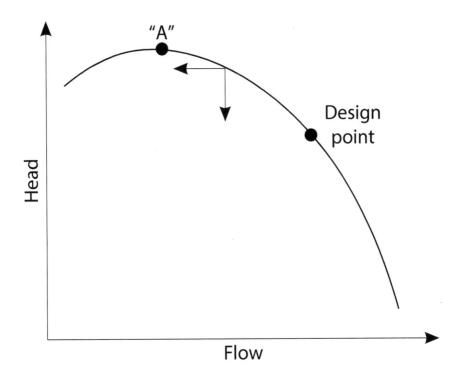

From plant data on the new giant vacuum tower residual pump. I plotted observed flow vs. pump discharge pressure!

Well! It's contrary to the manufacturer's curve! The head and flow are both going down, at a low flow. Impossible, I'd say!

The pump is never supposed to operate at such a low flow. But, Carl, you purchased this oversized pump yourself!

 Oh! Well then, I guess it's OK to run it below point "A", as we have lots of excess head anyway!

No, Carl! It's not OK. Below point "A", the pump vibrates in a most alarming manner!

 Constant criticism! That's all I ever hear from that woman! I suppose I might have added a spill-back in my design. Clare should have suggested that!

No need to mention this to Carl ... but Norm first saw this in 1991, at the Coastal Refinery in Aruba! Norm still complains to this day that those pump vibrations loosened the fillings in his teeth! Imagine what that did to the pump's mechanical seal? The pump had a minimum flow spill-back, but when Norm opened it, the pump lost suction and cavitated is a most alarming manner.

CONDENSATE BACK-UP IN CONDENSERS-THE EFFECT OF SUB-COOLING

Clare! Let's get condenser "B" cleaned. Look how high its outlet temperature is!

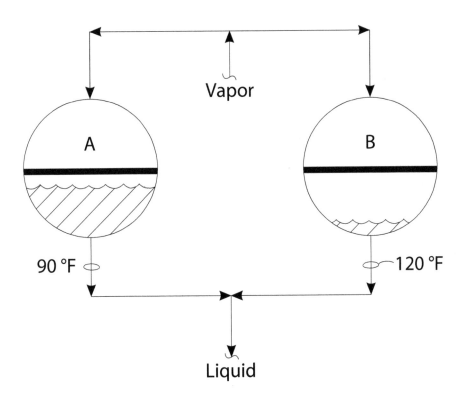

Actually Carl, it's "A" that's really underperforming!

 But the outlet flow of "B" is hotter than "A"?

Well, that's true! But only because "A" is suffering from CONDENSATE BACK-UP and sub-cooling! About 40% of the tubes in "A" are submerged in liquid, but only 10% of the tubes in "B" are covered in liquid!

 So, Clare, I suppose that now you have X-ray vision too? How can you know that "A" is suffering from condensate back-up?

Just run your hand along that channel head cover! You can feel the temperature gradient; where the channel head feels cooler, is the liquid level. Tubes covered in liquid, can't condense any vapors! Because the vapor is not even in contact with any of the tubes.

  Sometimes I think that Clare does have X-ray vision! She's like a Superwoman!

 You know Clare, this kind of reminds me about the condensate back-up problem we had in the channel head of our steam reboilers! Kinda the same process problem!

DISTILLATION TRAY DOWNCOMER BACK-UP AND LIQUID FLOODING

Clare, let me explain what causes downcomers to flood! It's because the downcomers are too small! That is, downcomer loading exceeds 175 GPM per square foot of downcomer cross sectional area!

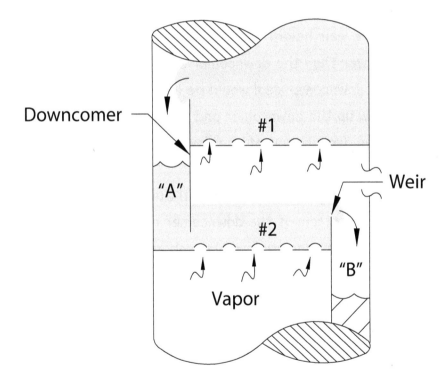

I certainly agree, Carl! But there are many other reasons for downcomer flooding. Shall I explain?

Why must this woman constantly correct and criticize me? We're not married!

Such as?

Well, Carl! Such as loss of the downcomer seal! If the weir height is adjusted wrong, it may be lower than the downcomer clearance! Then the downcomer seal would be lost! Vapor would blow up the downcomer and prevent liquid from draining out of the downcomer!

Easy enough to prevent! Just keep the bottom of the downcomer real close to the tray deck below. No problem!

No, Carl! Restricting the downcomer clearance will cause too much head loss under the downcomer, and also increase downcomer back-up! Also, if tray #1 gets too dirty, the flowing vapor pressure drop through tray #1 will go up. This will push up the liquid level in downcomer "A"! And Carl, if tray #2 leaks really bad, the bottom of downcomer "A" will

become UNSEALED! This would also cause vapor to blow up downcomer "A" and retard liquid drainage from tray #1!

I guess this would cause tray #1 to flood. That's bad!

Not only tray #1! With time, all the trays above would also flood!

∘ ○ ⟨ Why does that woman have to make everything so darn complicated? ⟩

OK, Clare! Just be careful, when you inspect the tray installations, that you have a $\frac{1}{2}$" overlap, between the top edge of the weir and the lower edge of the downcomer from the tray above! And make real sure the weir and bottom edge of the downcomer are LEVEL!

EFFECT OF FOAM ON LEVEL INDICATION IN DISTILLATION TOWERS

 Clare, I've just checked the level in the tower. It's 2 ft. below the reboiler return nozzle. It's fine!

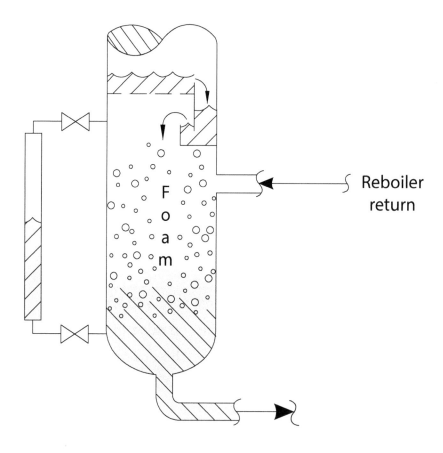

Reboiler return

Actually Carl, it's too high! I believe the high tower bottoms level is causing the tower to flood!

But can't you see, Clare, that the level in the gauge glass is quite low? Look at the level, woman! Use your eyes for a change!

But Carl, the level in the tower is higher than the level in the glass because the specific gravity of the liquid in the glass is greater than the specific gravity of the FOAM inside the tower!

How can you tell that? I don't see any foam in the glass!

It's simple, Carl. I lowered the level in the glass by 8" and the tower Delta P dropped by 20%! Understand now?

No! I'm totally confused!

Lowering the external liquid level by 8" may have reduced the internal tower level

(i.e., the foam level), by 16"! That is, below the reboiler return nozzle. This then stopped flooding of the trays, due to a high foam level! Norm calls this: FOAM INDUCED FLOODING! Do you remember Norm? He bought us lunch last month!

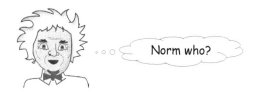

Norm who?

What we really need, then, in foaming services, is radiation type level measurement, which we could calibrate, for whatever foam density we wanted! Then we could know the real level for certain! I think the k-ray company sells a neutron back-scattering device to measure levels and foam densities.

SPLIT LIQUID LEVELS IN VERTICAL
VAPOR - LIQUID SEPARATORS

 Look, Clare! Split liquid levels indicate that we have a vapor and liquid sandwiched together! Liquid-vapor-liquid-vapor!

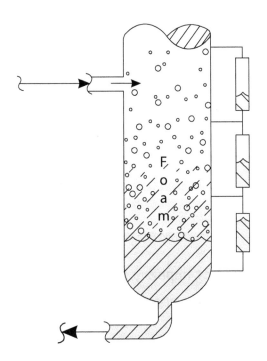

Not really, Carl! It's a sign of foam!

Foam? But why, dear woman, do we see three different levels, in the three glasses?

Because, Mr. Carl, the lighter, frothy foam, floats above the denser foam, which floats above the clear liquid. You can observe the effect in a glass of beer!

Then what does the level in the gauge glass represent? I'm confused!

The density of the foam between each level tap. If the gauge glass is 1/3 full, the foam density is 33% of the density of the clear liquid in the gauge glass!

But what happens Clare, if the foam rises above the vapor-liquid inlet nozzle?

MASSIVE ENTRAINMENT! TRAY DECK FLOODING!

For once I agree with Clare, based on my extensive beer drinking experiences!

OPTIMIZING EXCESS AIR IN A FIRED HEATER TO MINIMIZE FUEL CONSUMPTION

Oh, Clare! Please cut back on the heater air register! We have 4% O_2 in the flue gas, and our target is 2-1/2%!

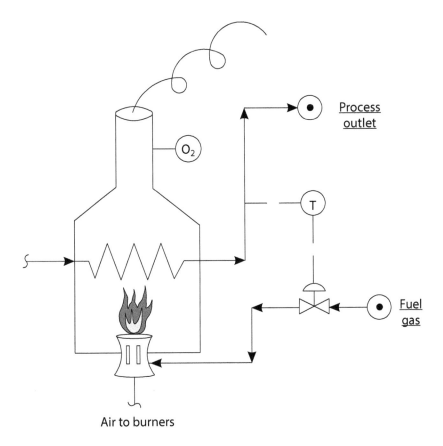

Air to burners

Not a good idea, Carl! The 4% is a minimum for safe operations for this heater, at this time!

But Clare, our heater expert says the optimum excess of O_2 is 2%-3%. Do you think you know more than the heater expert too? Don't be such a know-it-all.

The optimum O_2 is a function of burner air-fuel mixing efficiency! The burners are in bad condition, and their mixing efficiency is bad!

But still, we'll save some fuel gas, at the lower excess air?

Carl! If we fall below the optimum O_2, then the heater outlet temperature will go down, as fuel gas oxidation is suppressed, and fuel gas reduction and reactions are favored!

Oxidation? Reduction? What is that conceited woman talking about? I'm totally confused!

Oh, yes! Please elaborate, Clare!

Oxidation is an exothermic reaction, which releases heat. Reduction is an endothermic reaction, which absorbs heat. If we start to favor a reduction reaction because, in this case, we have too little air, the heater outlet temperature will be driven down!

But then, the fuel gas flow will increase, which will favor the endothermic reduction reaction more and more!

Quite true, boss! This will drive the fuel gas regulator valve open more and more! The heater will be trapped in a positive feed-back loop!

If that happens, why, we'll just open the burner air registers! We'll open the stack damper! You know, to get more air.

This may possibly blow the heater up! At Norm's former employer, two operators were killed by this mistake. It happened at their plant in Norco, Louisiana.

I wonder why we have O_2 ANALYZERS at all? I'd ask Clare, except she always gives me such complicated answers!

Hmm! I guess the optimum O_2 can only be found by trial and error. The optimum excess O_2, is that air flow that minimizes the fuel gas rate!

DISTILLATION TRAY DUMPING OR WEEPING WITH VALVE CAP TRAY DECKS

 Clare! Thank God for the Koch brothers! They invented the valve tray cap, which prevents tray deck leakage and weeping!

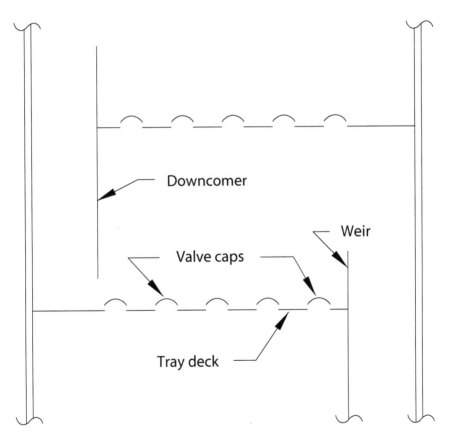

In reality, valve caps are a fraud! They don't work!

Watch your mouth, woman! That's the Koch family you're insulting!

Carl! I know you believe that at low vapor rates, the valve caps are supposed to act like little liquid check valves, and stop tray deck dumping.

Yes! That's the idea. Isn't this true?

No! FRI data shows that even when the tray decks are very level, valve trays are only marginally better than 3/8" hole sieve trays!

Then, Clare, what does keep valve trays from dumping, if not the valve caps? Don't forget trays have 2" - 3" high outlet weirs! There's a couple of inches of liquid on each tray deck!

Yes, Carl! Also, the CREST height contributes to the weight of the liquid on the tray decks! It's the Delta P DRY that prevents tray deck dumping and weeping!

Delta P dry? Clare, what does that mean? I'm getting confused!

It's the pressure drop of the vapor, as the vapor accelerates through the perforations on the tray deck! Delta P dry should be about equal to the weir height plus the crest height, taking into account tray deck levelness, plus the tray aeration factor (typically 50%).

What is wrong with that woman? Doesn't she know that the Koch family controls the destiny of America? Next thing she'll advocate is the use of ancient tunnel cap or bubble cap trays to improve tray efficiency by retarding tray deck weeping or leaking.

FIRED HEATER - TUBE FAILURES

 In general, Clare, furnace heater tubes fail due to localized overheating. This melts the tubes!

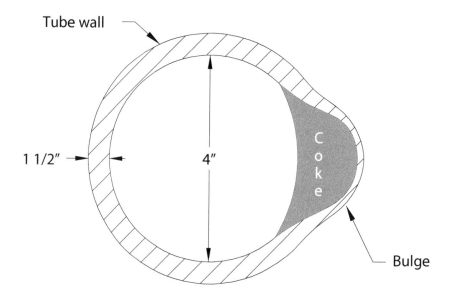

Tube wall

1 1/2"

4"

Coke

Bulge

That's not quite right, Carl. The tubes don't melt. They undergo HIGH TEMPERATURE CREEP!

Creep? What's that?

Oh Carl! At 1380° F, high chrome tube walls become deformable. Then, the internal pressure causes the tube wall to bulge out!

I see! It's like blowing up a balloon! The wall of the balloon gets thinner and thinner, as the balloon expands!

That's right! The tube wall thickness at the bulge becomes too thin to constrain the tube internal pressure of the process fluid!

But what causes the local hot spot? Localized overheating? Flame impingement?

Not usually! Most often, it's coke formation inside the tube, due to low mass velocity (less than 100 lbs / ft² / second), or feed interruptions, or left over coke from prior decoking operations during a turnaround! Velocity steam in the heater passes also helps protect the tubes from failure!

I hope all our refinery friends will remember to keep their box SNUFFING STEAM lines operable, in case they have a tube leak! The alternative to snuffing or box purge steam is called death!

You're right, Clare! But for Delayed Cokers, it's best to keep a minimum velocity of 200 lbs / ft² / second! But you're talking about coil steam, not snuffing steam.

LOW AIR FLOW IN A FIN FAN FORCED DRAFT AIR COOLER

Clare! Check the pressure drop of the air flowing across the fin fan air cooler bundle! I think the bundle is fouled. The fins are full of dead moths!

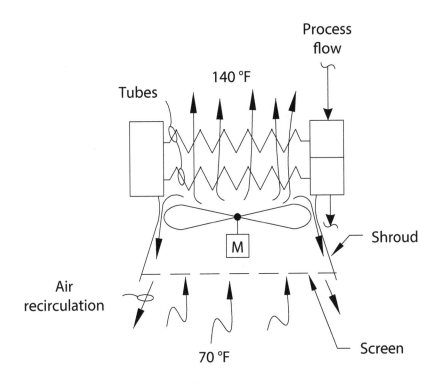

Process flow

140 °F

Tubes

Shroud

Air recirculation

70 °F

Screen

Carl! That won't serve any purpose! The fan discharge air pressure and discharge flow remains relatively the same, even when the outside of the fins get fouled!

But look how hot, 140°F, the air blowing out of the top of the fins is running! Certainly we're getting less air flow and more delta P?

The air flow is less through the bundle, as you've said! But the flow leaving the fan is constant! The air flow through the bundle just drops off to keep the DP constant!

Well I'm confused. It seems like some air has gone missing, if the flow from the fan is constant, but the flow through the tube bundle drops off?

Your "missing" air flow is recirculated around the periphery of the screen, underneath the fan. A small piece of paper will be blown off the screen around its edge.

That's true! But I always see this recirculation flow!

Yes! But as the fins foul, the amount of recirculated air increases, and the amount of air flowing through the fins decreases, but the total air flow is approximately constant!

MEASURING AIR FLOW FOR AN AERIAL FIN FAN AERIAL COOLER

 Go get an anemometer! That's used to measure air velocity. Clare – wind speed. We need to calculate the air flow through the aerial cooler!

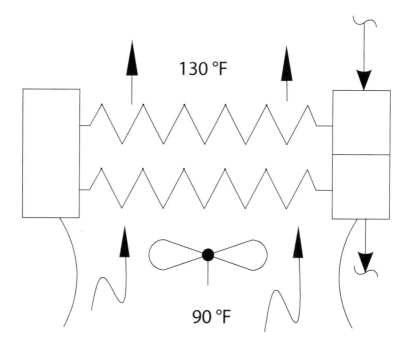

130 °F

90 °F

Actually, Carl, that's not the way it's done!

What is the way? I think the fan is not spinning fast enough and air flow is too low! Maybe the belts are slipping?

Well, Carl, I've measured the air inlet and outlet temperature (130°F - 90°F = 40°F) to get a temperature rise of 40°F. The specific heat of air is 0.25 BTU/LB/°F. Thus, each one pound of air is picking up 40°F x 0.25 = 10 BTU/LB.

Okay! But how do we calculate the air flow?

I need to calculate the process duty. Let's say that's 10,000,000 BTU/HR.

The air flow is then: 10,000,000 BTU/HR ÷ 10 BTU/LB = 1,000,000 LB/HR of Air Flow

That's not the way I learned to do it at my university.

Thanks, Clare! That seems simple enough! We can compare the calculated result to the vendor's performance data sheet!

Yes! But the vendors never take into account air recirculation, which happens even when an air cooler is new and clean. Therefore, the observed air flow through the bundle is always too low by maybe 20%.

MEASURING COOLING TOWER EFFICIENCY
APPROACH TO WET BULB TEMPERATURE

 Clare! There's something wrong with the cooling tower!

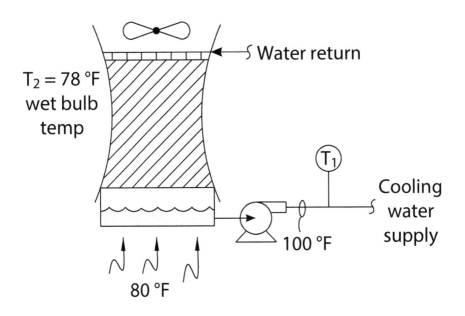

 Mr. Carl! Why do you think the cooling tower performance is bad?

Use your brain, woman! The ambient air temperature is 80°F, and the effluent water from the cooling tower is 100°F. That's a 20°F delta T. The unit was designed for a 12°F.

True enough, boss! But the cooling water tower was designed assuming very low relative humidity. This is New Orleans, and the humidity is really high!

Relative humidity? I've heard about that! How does that affect the cooling water temperature?

First, I'll measure the WET BULB TEMPERATURE! I'll tie a piece of cloth around the bulb of a lab thermometer. Soak it with water and then swing it around for a minute. This gives me the wet bulb temperature.

That sounds familiar! Then what?

Next, subtract the wet bulb temperature
(78°F) from the cooling water return
temperature (100°F), to obtain the approach
to the wet bulb (100°F - 78°F) = 22°F. An
approach temperature of:

- Less than 10°F – excellent
- Above 20°F – terrible

So, admit it, Clare! I was right! The cooling
tower is defective!

Not exactly, sir! It looks like the belt is
slipping on the cooling tower's induced draft
fan! That's pretty easy to fix!

Good Lord! Even when I'm right, Clare will
never, ever give me any credit! She always,
constantly, has to "one-up" me! Why can't
she be more like my mother? Mom
knew how very smart I was!

Clare! It could also be that the angle of the
cooling tower's fan blades are not set steep
enough. I read that $22\frac{1}{2}$ degrees is pretty
much the optimum.

ADJUSTING HEATER STACK DAMPER FOR OPTIMUM ENERGY EFFICIENCY

 Clare! Make sure that the heater stack damper is half closed, then you may adjust excess air using the air registers, at the base of the heater!

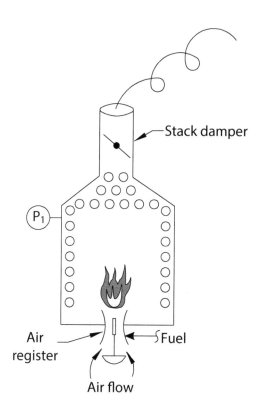

Perhaps that's not a good idea, Carl! We could develop a positive pressure at P_1, below the bottom row of convective section tubes!

Okay! That could damage the heater's roof arch supports. Better to open the stack damper 100%!

Perhaps, Carl, that's not a good idea either! Too low a pressure at P_1, will draw cold tramp air into the convective section, cool the flue gasses, and reduce convective section heat recovery!

Hmm? Then just partly close that stack damper until the pressure at P_1 is about 0.05" to 0.10" of water draft. I read in Norm's books that's the optimum draft.

What Norm forgot about was wind! If the wind dies off suddenly, the draft may suddenly decrease, and the pressure at P_1 would go positive. Then, if you're burning high sulfur fuel, the SO_2 will blow out of the fire box and could easily kill someone! In New Orleans, it's not so windy, and Norm forgot about this danger!

No matter what I say, that woman always has to contradict and criticize me!

Okay, Clare! Do your best! I've got to go to an important meeting now! I agree that the air registers and the stack damper need to be used as a team to optimize excess air and draft simultaneously.

PREVENTING TRAY DUMPING BY USE OF BUBBLE CAP TRAYS

 Clare! Bubble cap trays should never be used in a modern design of a distillation tower! Use modern grid trays!

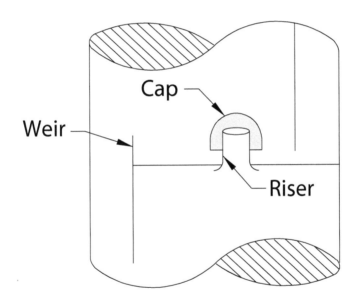

Bubble cap tray
(only one cap shown)

Sorry, Carl! But I tend to disagree. Bubble cap trays do have distinct advantages!

No, Clare! Don't you realize that bubble cap trays have 10-15% less vapor handling capacity, before jet flooding, than do valve, or grid, or even sieve trays?

That's very true, Carl! But bubble cap trays are not subject to tray deck dumping, or leaking, or weeping! As long as the height of the riser is above the weir!

Hmm? But why are bubble cap trays not widely used any longer? Even a modern grid tray will lose tray efficiency when the vapor pressure drop gets too low, and the tray deck dumps liquid. That promotes vapor-liquid channeling!

Mainly, boss, because during turnarounds, it's a lot of trouble to remove each bubble cap, and clean between the riser and the cap!

That Clare has a lot of old-fashioned ideas! But I still like that old gal!

You're right, Clare! Bubble cap trays will have a better turndown efficiency than do perforated tray decks. Let's use some! Especially in distillation services where the vapor flow rate can vary a lot with feed and reflux rates.

DEMISTER FOULING IN VAPOR-LIQUID SEPERATOR VESSELS

 Clare! Use of a 4" thick stainless mesh demister pad in a vapor-liquid separator is always good design practice.

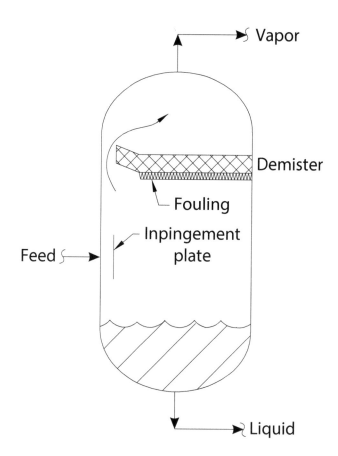

Not always, Carl! Sometimes a demister will promote entrainment!

Don't be a fool! Demisters cause entrained droplets to coalesce into larger droplets and settle out more readily! Haven't you even heard of Stoke's Law?

But Carl, in refineries, demister pads will foul and partly plug. Then delta P across the demister will cause it to be torn from the vessel wall. Vapor channeling and locally high velocity will then cause worse entrainment!

Drat that woman! She's always right lately!

Okay then! We'll just use demisters in clean services – which I guess precludes their use in most refinery services.

Actually Carl, what works better than demisters is an IMPINGEMENT PLATE! If designed correctly, it converts one radial entry point into dual tangential entries. Norm tells me he uses this design, especially in towers with high inlet velocities. The idea is to split the flow as it enters the tower in two opposite directions around the inner circumference of the vessel.

For real bad entrainment, a "Vortex Tube Cluster," will do a super good job of separating vapor from the droplets of liquid.

EFFECT OF TEMPERATURE ON LIQUID LEVEL INDICATION

Mr. Carl! You're running the tower bottom level too high!

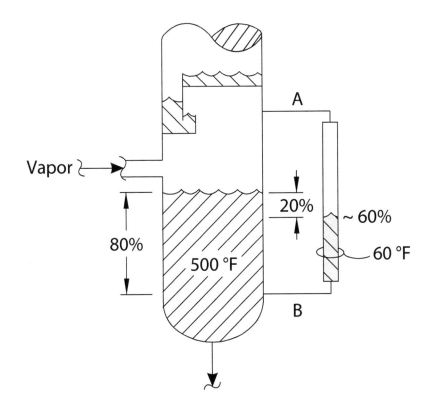

Clare! I know what I'm doing! The level is only 60%. Use your eyes, woman! Look at the level glass!

But Carl, the glass is 500°F colder than
the tower. The s.g. of hydrocarbon liquids,
increases 5%, for each reduction of 100°F:

$$\bullet \frac{(560°F - 60°F)}{100°F} \cdot \bullet 5\% = 25\%$$

What's your point, Clare? We're talking about
levels, not density.

Dear man! Do not yell at me! The liquid in the
glass is measuring the pressure difference
between points A and B, in terms of the s.g.
of the fluid in the glass! As the s.g. of the
hotter liquid is 25% less than the 60°F liquid,
the tower level will be 25% higher than the
level in the glass!

Hmm....? I guess that means that if the level
in the glass is like 60%, then the level in
tower might be 80%! Is that right, Clare?

At least, boss! If there is **FOAM** inside the
tower, the discrepancy may be much worse! If
the bottom level rises above the vapor inlet,
the liquid or foam in the bottom of the tower
will be pushed up against the bottom tray!
Then the entire tower is going to flood!

This reminds me of what happened at Texas City (BP) on their giant Naphtha Splitter, when all those people were killed about ten years ago. Seems like Norm helped design that tower in the 1970s. Those guys should have connected the pressure relief valve to the flare, **not** to an atmospheric vent. In that case, the bottoms product was less dense than normal, so the panel level indicator was reading too low! Same idea, I think? I don't believe though that I'll ask Clare about it! She'll just make another long speech!

DRAW-OFF NOZZLE CAPACITY LIMITS

 Clare! Forget about your project about adding a new nozzle on the crude tower!

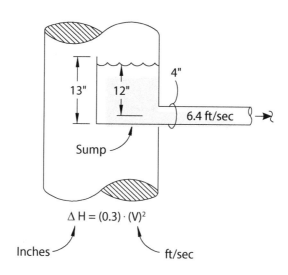

$$\Delta H = (0.3) \cdot (V)^2$$

Inches   ft/sec

But Carl, the existing 4" nozzle is too small. The hydraulic head loss through the nozzle is 12":

And the height of the sump above the center line of the nozzle is only 13"! Without a new 6" nozzle, the sump will overflow and dump kerosene into diesel!

Pay attention! The code stamp on the vessel says "Stress Relieved – Do Not Weld"! What part of that, Clare, don't you understand?

Sorry, Carl! That just means that if we do weld a new nozzle on the tower, that the weld has to be Post Weld Heat Treated or stress relieved, to stay within ASME boiler code requirements! It's no big deal – maybe $10,000?

Sounds like a lot of trouble. But as an American, I always support my local ASME!

I don't know why we can't just raise pressure to force more kerosene out of the tower. I suppose Clare will tell me we can't, based on her woman's intuition! Or, some bubble point composition of the kerosene nonsense?

ON-STREAM REPAIR OF TUBE LEAKS IN SURFACE CONDENSER

Oh Clare! Please tell operations to shut down the main air blower on the FCU! We've developed a giant tube leak in the surface condenser! There go our profits for the 3rd quarter!

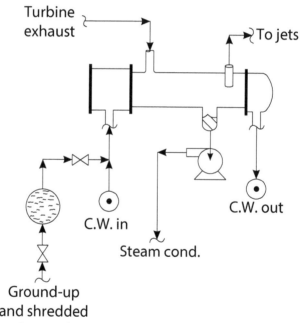

How do we know that, Carl? That's terrible news!

Dear girl! Don't you ever look at the lab results? The conductivity of the steam condensate is through the roof!

You're right, Carl! But we can maybe fix the leak on-stream!

 On-stream? Like with magic? Wake up to the real world!

No, sir! We can inject shredded fiber glass into the cooling water supply, provided the leaks are not too gigantic and have developed gradually. Norm used to use asbestos fibers in the 1960s! It's a way of buying time.

 Sounds simple! But is there a downside? Don't forget what that French woman said, "You can't have your cake, and eat it too!" Ha-ha! Personally, I don't like cake!

Well, yes! The problem could be the fibers could plug up the pump's suction screen. Our vacuum condensate pump has external seal flush, so the seal is not a problem! Also, we would need to put the condensate to the sewer, to avoid returning hardness deposits to boiler feed water! Also, Marie Antoinette was Austrian, not French!

 Okay! Let's give her a try! If it works, we won't have to shut down the CAT!

 If this really works, I'll tell Mr. Felder it was my idea! If it doesn't, I'll explain it was that woman's fault! Ha-ha, ha-ha!

WHEN ARE VORTEX BREAKERS REQUIRED?

It's good engineering practice to use a VORTEX BREAKER in all draw-off nozzles! The vortex breaker should be shaped like a cross! Can you remember to follow this accepted standard, Clare, in all your designs?

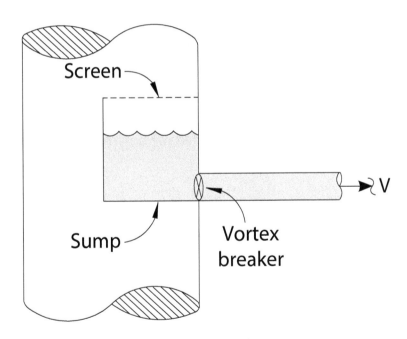

Screen

Sump

Vortex breaker

V

I don't totally agree with you on this, Carl!

What now, woman? Why do you always want to debate everything?

Well, for one thing, at velocities of less than 2 ft/sec, vortexing in nozzles is not a problem! At draw-off nozzle velocities above 4 ft/sec, vortex is sure to be a problem!

And how do you happen to know that, Clare?

I performed some experiments in my bathtub! I like to bathe!

Still, it seems like we should use vortex breakers all the time, as they're cheap and don't hurt anything. Better safe than sorry!

Actually, vortex breakers can cause a great deal of trouble. Stray pieces of tower internals and turnaround trash may be caught up in the vortex breaker vanes and restrict flow!

Hmm! I hadn't thought about that. But then, I take showers! Ha! Ha! It's quicker than a bath! But what if nozzle exit velocities are above 4 ft/sec? Then, you have to use my idea of a vortex breaker!

You're quite correct, Carl! But then Norm suggests a 1" x 1" mesh, 316SS screen be welded across the top of the sump, to protect the vortex breaker from plugging!

Well, Clare, okay! I don't have time to worry about your little details!

Clare worries too much! She should always look on the bright side of life! If she's so concerned about vortex plugging, she should issue instructions to maintenance to clear up and properly secure tower internals before closing a vessel during the turnaround!

NAPHTHA INJECTION TO CENTRIFUGAL COMPRESSOR

 Look here, Clare! Have you utterly lost your brains? You can't inject liquid into a compressor! You'll tear up the rotor and break the valves!

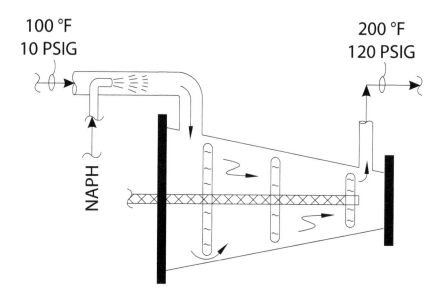

Correct, boss! For a recip, one should never permit liquids to enter the compression cylinders! However, for a centrifugal compressor, a finely divided mist or spray can be helpful!

And why is that, dear woman?

Well, the gas enters the centrifugal compressor at its dew point temperature and pressure. And then, as it's compressed, the heat of compression superheats the gas. The gas dries out, and any solid fouling deposits, like salts, precipitate on the rotor's wheels! It's a slow, but steady process!

Hmmm? I was just wondering, if you're so smart today, why are the amps on the motor driver of the coker wet gas compressor going down, as the rotor fouls? You'd think the amps would go up?

Fouling causes a large reduction in compressor capacity. The efficiency drops too, but the capacity declines faster.

Well, Miss Brilliance! What are we going to do about the rotor fouling on the coker, hydrocracker, and reformer centrifugal compressors?

Carl, as I tried to explain a moment ago, spraying a finely divided mist through a spray nozzle will prevent the gas from drying out and keep the rotor clean and will...!

...and probably wreck the machine!

No, Carl! Not if the spray injection rate is a finely divided naphtha mist calculated to absorb the heat of compression by evaporation! Also, the liquid spray has to be tripped off when the compressor shuts down!

Well! Go ahead with your calculations! I've got an important meeting to attend!

I've got to talk to Mr. Ratbone, the Human Resources Manager! I need to make sure I'm getting full credit for having a female on my staff. I ought to get some extra special credit for putting up with that Clare!

 Hmmm? I suppose it would be really, really bad to spray a mist into the suction of a reciprocating compression? Likely wind up wreaking the discharge valve plates!

INTERNAL OVERFLOW FROM TOTAL TRAP-OUT CHIMNEY TRAY

Oh, Clare! That FCU Debutanizer is flooding! We're going to have to cut cat feed! I guess the trays are fouled again. Let's plan a shut-down to chemically clean that rotten tower again!

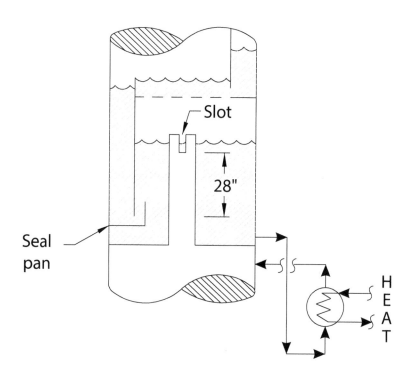

Sorry, Carl! That's not our problem! We cleaned that tower six months ago – no help! It's not dirty trays!

Look, Clare! Don't argue! Just for once, can't you follow my instructions? UQP reviewed the tray capacity! They're only 68% of jet flood and 45% of downcomer back-up! Doesn't that prove that the trays are fouled? Think, woman! Look at the Aspen simulation!

Actually, Mr. Bossman, it's a UQP design error that caused the problem! The reboiler shell side has become fouled, and its shell side DP has increased a few psi. That pushes up the liquid level in the chimney tray feeding the reboiler! As there's no adequate provision for internal overflow, the high liquid level in the chimney tray causes the bottom downcomer to back-up and flood!

So now, Clare, you think you're also smarter than UQP! Well, you're not! UQP included a 6" slot for internal overflow in the chimney! I've seen that slot myself!

Yes, Carl! But did you happen to also see that the bottom of the slot is 28" above the seal pan overflow lip? This will lead to excessive downcomer back-up and flooding! We need an 8" overflow pipe, with its top edge in-line or below the seal pan!

 That's an awful way, Clare, to talk about UQP after that party they made for us last Christmas! How many crab cakes did you stuff yourself with?

 Gee! You've got to look at every tiny detail! I need to get off this technical track and into upper management! Maybe Human Resources?

VACUUM EJECTOR – LOOSE STEAM NOZZLE

Clare! Exactly why is it, that you have partly closed the motive steam nozzle to the first stage ejector? Is this one of your misguided attempts to save steam? We need to be maximizing vacuum, not saving steam!

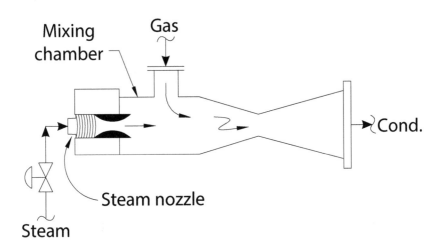

Yes! I'm quite aware we should be maximizing vacuum! That's why I have optimized the jet's motive steam pressure!

But, woman! The manufacturer's design motive steam pressure is 150 psig! Look at the name plate! Can't you see it says "150 psig"? I suppose you're smarter than the ejector manufacturer now? Well?! I'm waiting for an explanation!

All I can say, Carl, is that I reduced the motive steam pressure to 118 psig, and the vacuum improved from 42 to 35 mm Hg. The usual reason for this is that the 316 SS nozzle has become loose where it screws into the carbon steel body of the jet! And the motive steam is partly blowing into the jet mixing chamber, but not through the nozzle!

Hmm...? I wonder why the manufacturer didn't make the back part of the ejector out of 316 stainless also? You know, Clare, to avoid galvanic corrosion in the presence of wet steam? I guess they have a really good reason?

Or maybe, the nozzle is partly eroded? Or maybe, the downstream condenser is fouled and can't condense the motive steam completely? It's kinda hard to know, Carl!

Okay, I get your point! Let's throttle-back a bit more on the motive steam and...! Oh no! The jet's surging! The vacuum's breaking! Look, the pressure's jumping up! Help, Clare! Help! Do something, quick!

Carl! You've forced the jet out of its "CRITICAL MODE," of operation! The jet has lost its sonic boost! You've got to be more careful when you try to optimize the motive steam pressure! Take your time! Reduce the steam pressure slowly and cautiously! Don't rush!

All right, Clare! A little less criticism would be appreciated! I've got it!

That woman is always too careful! Sometimes, men just have to move boldly ahead in life! Clare doesn't understand male boldness! It's a macho-type thing!

EFFECT OF TRAMP AIR LEAKS ON HEATER EFFICIENCY

 Next time you're outside, Clare, pinch-back on the heater air! We've got too much O_2! Either on the stack damper or air register! It doesn't make any difference! But don't forget!

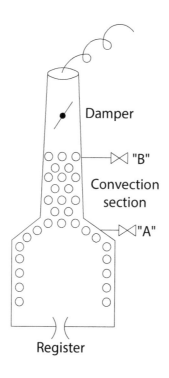

Damper

"B"

Convection section

"A"

Register

It surely does make a difference! If I close back on the air register, draft will go up at Point "A"! If I close back on the stack damper, then draft will go down. Too much draft is bad. And too little draft can result in a positive box pressure, if the wind slows later on!

Okay, okay! I sure don't want a positive box pressure! That's dangerous! So just close those burner secondary air registers halfway, Clare! Understand? Okay?

Not okay! If we develop too much draft, then cold tramp air will be sucked into the leaky convective section! Then, Delta O_2 will go up! That's the difference between the O_2 at Point "B" minus Point "A". Then, the percent of the heater fuel wasted, due to the cold tramp air, quenching the convective section hot flue gas, will be:

$$\bullet \text{ \% Fuel Wasted} = \frac{(\text{Delta } O_2) \bullet (\text{Delta T})}{500}$$

Where Delta T is the stack temperature, minus ambient air temperature, in °F! (Use 300 if working in °C)

So, if I want to cut back on combustion air, if I close the air register, I'll suck in too much tramp air leaks! But if I close the stack damper, then the radiant firebox pressure might go positive! And, blow-out hot flue gasses with deadly SO_2 out the sight ports! My, oh my! Life sure is complicated!

EFFECT OF A SINGLE FOULED TRAY

I just inspected the crude unit debutanizer, Carl! Big problem! We'll have to get Mr. Thompson in maintenance to hydro blast tray #20. It's fouled, and those tray caps are sticking to the tray deck!

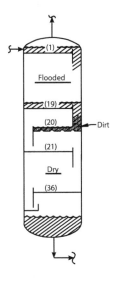

No! I've approved the debutanizer for closure yesterday! Yeah! I saw that feed tray – tray #20 you say? – was pretty fouled with gunk! But the trays above and below that feed tray were clean! It's just that single tray #20 was missed, and didn't get hydro blasted clean. I'm not gonna get those maintenance guys mad at me for just one tray! You gotta learn to be diplomatic, Clare! Don't wanna get Tommy mad at us!

Carl! If the valve caps are sticking on one tray deck, then that tray will develop a high pressure drop! It would be better to break off the valve caps, then start up with those caps sticking to the floor of the tray!

Listen to me, woman! I don't care 'bout a little extra delta P on the debut! It ain't no big problem! How 'bout makin' us a new pot of coffee, and stop worrying 'bout nothin'? Stop complaining!

Make your own damn coffee! It's not the pressure drop! It's that a high vapor delta P on one tray will cause that tray to flood due to downcomer back-up. And then, all the trays above will flood with time! And all the trays below will dry out! And flooded trays and dried out trays don't fractionate!

Clare – you been readin' that Lieberman book again – *A Working Guide to Process Equipment*? Is that where you're gettin' these dumb ideas from? Okay. I don't like to argue with women! Call Tommy on your radio. Channel 6. And get maintenance back down here to clean that tray #20. Will that make you happy?

I reckon Clare's just a perfectionist. Needs to get everything perfectly clean! Even washes out the coffee pot each time – which weakens the next pot of coffee!

STEAM TURBINE – SURFACE CONDENSER OUTLET TEMPERATURE

We've just received instructions from Tech Service to reduce the surface condenser outlet temperature! Their idea, Clare, is that a lower surface condenser outlet temperature, will reduce the turbine exhaust steam pressure! This allows more horsepower to be generated from each pound of steam. We might be able to save 5% of the steam to the K-805 steam turbine, if we can knock 5°F-10°F off the surface condenser outlet!

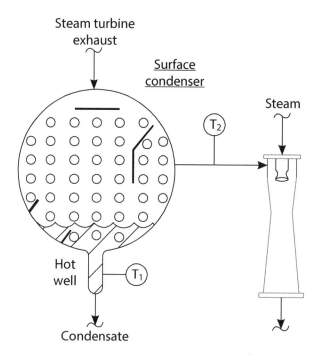

Yes! Quite so! I believe Norm calls this an improved "Isentropic Steam Expansion." The lower the condensing steam temperature, the lower the surface condenser pressure. I'll try to get the surface condenser temperature cooler by turning on another cooling water pump and also back-flushing the condenser's tubes!

Why is it you always want to do things the hard way? All you got to do, Clare, is just raise the level in the Hot Well! Then you'll see the boot temperature will go down! Look at the dial thermometer at T-1! See, that's the reason we got experienced men like me out here! Understand now, Clare? I got me 22 years of experience out here!

Yeah! You've got one year of experience 22 times! Raising the Hot Well level too much will back-up condensate over the tubes! The condensate is gonna get sub-cooled, and lots colder! But the number of tubes exposed to the condensing steam will be reduced. Then the vapor outlet temperature at T-2 will increase!

Who cares? We monitor the temperature in the Hot Well, Not the vapor outlet! The vapor temperature is not important! Don't you read the operating manual? It says to keep the Hot Well cool! Can't you read, Clare?

The turbine exhaust pressure is controlled by the vapor pressure of water at the condenser **VAPOR** outlet! That means T_2 – not T_1! If you back-up the hot well, the pressure in the surface condenser will increase and the turbine is going to slow down with the same amount of driver steam flow! Try it and you'll see!

How I hate that woman! I raised up that boot level to 100+%, and it slowed down the turbine, at a higher steam flow! It seems lately like she's right about everything! Oh well! In another 18 years, I can put in for early retirement! I'll still only be 62 years old!

Later that Day

 Oh, Clare – would you mind starting P-303-C next time you go out! Maybe a bit more cooling water flow to the turbine might help us a bit! Can I get you a coffee?

WATER ACCUMULATION IN TURBINE CASE

Uh-oh! Clare! That turbine is starting to vibrate again! It's water accumulating in the turbine case, hitting the wheels. If only the exhaust steam outlet was on the bottom of the turbine case! I suppose the drain on the bottom of the turbine is still plugged!

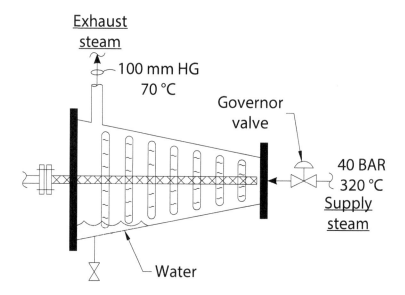

Exhaust
steam

100 mm HG
70 °C

Governor
valve

40 BAR
320 °C
Supply
steam

Water

Carl, that drain won't work anyway! That end of the turbine is under vacuum. We'll just have to dry out the exhaust steam! I'll start slowing down the turbine by closing off on the governor valve! That'll heat up the exhaust steam by about 10°C!!

No, woman! When you close off on the steam inlet valve, meaning the governor speed controller, the supply steam gets colder! Even you ought to know that much! Understand, Clare?

Do not use that tone of voice when you speak to me! I'm not your wife! Anyway, you're right, reducing the steam pressure will cause the steam leaving the governor valve to cool off. But throttling on the governor will heat up the turbine exhaust steam and make it drier!

Hmm... Yeah! I kind of noticed that the turbine exhaust steam temperature to our surface condenser does heat up when I slow that turbine down 50 or 60 rpm's. Seemed kinda strange! I don't know if...

If you look at the Mollier Diagram Norm gave us in his seminar last May, you'll understand! Throttling on the motive steam supply to the turbine is reducing the amount of horsepower, or work, that the steam supplies to the turbine wheels. That leaves more heat in the steam exhaust!

Yeah! It makes the turbine work less efficient. Which then heats up the turbine exhaust temperature! Seems backwards though to cool off the motive steam to heat up the exhaust steam! But I guess anything that works less efficiently starts to heat up! I noticed, Clare, that you heat up too, once in a while! Ha-ha!

Save those sexist comments for that poor woman who had the misfortune to marry you! It's just not a good design to have the turbine exhaust nozzle below the vacuum surface condenser, in a condensing steam turbine.

Yeah! Those dumb engineers should have elevated that there turbine high enough, so that we could drain right into that surface condenser. Uh... you got my coffee ready yet?

Norm's Mollier Diagram? What did I do with it? Clare, and Lieberman, and that Mollier guy, and the... what do all these people have against me?

CHAPTER TWO

CRUDE DISTILLATION

PROFESSOR POT & KUMAR

 I am Professor R.D. Pot! Head of the Chemical Engineering Department at M.I.T. (Mumbai Institute of Technology). I am also a famous consultant for Axxoco Oil.

Good afternoon; I'm Kumar!

Everything I know, I have learned at university from Professor Pot! I'm a Process Engineer at the Crude Unit at the Axxoco Refinery in Bridge City, Texas.

HOW TO ADJUST PUMPAROUND FLOWS

 Listen Kumar, my dear friend! Always keep the heavy straight run pumparound (P/A) at its maximum flow so as to maximize crude preheat!

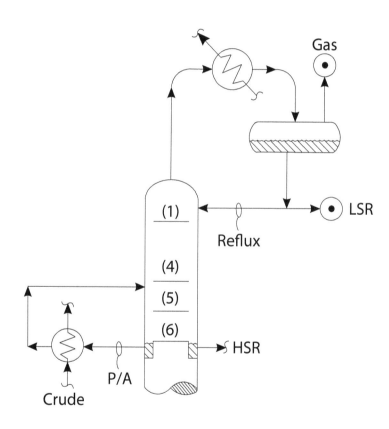

 Is that always true, Professor Pot? How will that affect fractionation between LSR & HSR?

Well, I hadn't thought about that. It seems like taking more heat out in the P/A, will reduce the top reflux rate and make fractionation worse between the HSR & LSR!

But Professor, if the top trays are dirty, and are running too close to their flood point, then reducing the top P/A duty will make them flood and really hurt the split between HSR and LSR, due to flooding and excessive entrainment!

Kumar! You are talking in riddles! Make up your mind! Are you for or against more top P/A heat extraction?

Professor, if we reduce the P/A duty, that may overload the top vapor condenser and make too much gas!

Yes! Yes, Kumar! That's why I told you to maximize the top P/A in the first place! To reduce overhead condenser load and to save energy, by maximizing the crude preheat, thus reducing the crude heater firing rate!

All true, Sir! But then the LSR 95% point will get too high due to the lower reflux ratio?

Apparently, the subject of optimizing the P/A circulation rate is very, very complex. And we are only discussing one of the three pumparounds on the crude column!

Yes, Sir! May I suggest we ask Norm's opinion on this subject? He worked on #12 Pipestill in Whiting, Indiana, for American Oil in 1965. After 50 years, he must understand how to adjust the pumparounds on crude distillation fractionators!

HOW TOP REFLUX RATE AFFECTS FLOODING ON TOP TRAYS

Kumar, reducing the tower top tray temperature with more reflux will reduce the volume of vapor flowing to the top tray of the crude distillation tower and stop the tray from flooding!

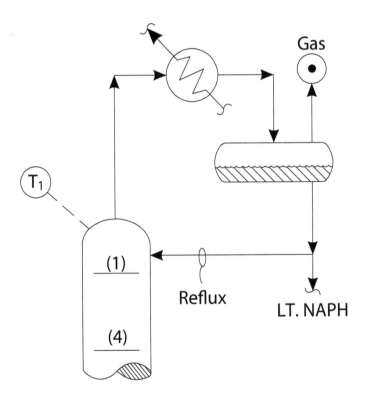

Are you sure, Professor Pot? It seems like the tower floods if we just raise the top reflux and leave all our pumparounds the same?

Listen to me! I'm a tenured professor here at M.I.T.! Raising top reflux cools off the vapors and shrinks them. Use your common sense!

Please pardon me, Professor! What I see in the plant is that raising the top reflux causes the pressure drop, DP, across the top four trays to increase! Then, if we don't do something, like increasing the diesel pumparound, we will begin filling up the reflux drum!

Hmmm! Those are symptoms of jet flood! Perhaps the trays...?

Yes, Professor! Yes! The big problem is that raising the reflux rate reduces the molecular weight and thus increases the mole per hour of vapor flow!

Well, that's true! More moles, more vapor volume per pound of vapor! But certainly the pounds of vapor flow decreases, as the vapors are cooled by the top reflux?

Actually, sir, the weight flow of vapor increases, as the tower top temperature cools, due to more reflux rate!

More mass flow as we lower the temperature at T1? That makes no sense Kumar! Where did you get such an idea from? Certainly not at M.I.T.!

I'm confused about this myself, Professor Pot! I will ask Norm to explain this to me.

Later that Day

Professor, Norm said that when you lower the top few trays' temperature, by increasing the reflux rate, the sensible heat content of the up flowing vapors would decrease because the vapors are colder. This sensible heat is absorbed by the liquid on the trays, which is saturated liquid at its bubble point. As this liquid absorbs heat from the vapor, it vaporizes to a greater extent. In effect, we are converting the sensible heat content of the vapors, to latent heat of vaporization of the reflux!

Yes, Kumar! And the reduced molecular weight multiplies the effect of increased vapor mass flow rates! I see that!

DESALTER - ADJUSTING MIX VALVE PRESSURE DROP

 Kumar! You were always my favorite student! Very serious! May I tell you something important about your crude unit desalter operation?

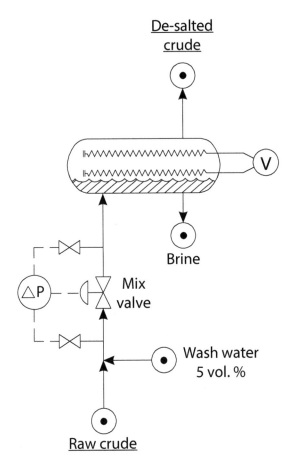

De-salted crude

Brine

Mix valve

Wash water 5 vol. %

Raw crude

Certainly, Professor Pot! I always tried my best at M.I.T.

Here are the desalter rules:

1. Use 6% volume wash water
2. Avoid increasing amps on desalter grid
3. Maintain a mix valve delta P of 10 PSI, plus or minus 20%

Hmmm, Professor…? A question? If we have emulsion problems and are carrying over a high BS & W, (1/2 wt% water) in desalted crude, perhaps added mix valve DP is bad? Maybe it would be best to open the mix valve completely? No sense having good mixing and carrying over emulsion, I think?

No! No! dear student! We must always follow the desalter rules, published by the respected chemical vendors! From where do you get such ideas? No desalter mix valve delta P—for shame! For shame!

I apologize most sincerely! But Professor, Norm suggested that not only should we leave the desalter mix valve 100% open, but that it should be removed from the unit and that it'd be…!

What I've observed, Professor, is that increasing the desalter temperature will always increase the desalter amps. And, on heavier crudes (20° API), I rather like to have a desalter temperature about 270°F - 290°F. Less viscosity! Of course, the higher temperature increases the conductivity of the crude. And it increases the solubility of water! But as long as the grid does not arc, and voltage is stable — I'm happy!

I'm afraid Norm is a bad influence on Kumar! Encouraging him to flout rules! What's next?

Yes! But don't forget that extra dissolved water due to higher crude temperature, will increase the water partial pressure in the crude tower overhead vapors! You don't want to fall below the water dew point temperature at the top of the crude tower, Kumar! This would cause corrosion and fouling on the upper tray decks.

CAUSES OF TRAY DECK FOULING

Kumar! Dirty tray decks will cause the valve caps to stick to the tray decks! This causes high vapor delta P. Which in turn, prevents the downcomers from draining! Typically, the flooding backs up the tower, from the tray where flooding is initiated. Or, we say, that flooding progresses up a tower, but not down!

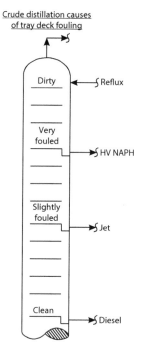

Crude distillation causes
of tray deck fouling

Thank you so much, Professor Pot! What then are the indications of crude tower flooding?

Ah, Kumar! What a blessing to have an excellent student! The indications of crude tower flooding; I will list for you:

1. Increasing the top reflux rate, increases tower top temperature, rather than decreasing it.
2. The heavy naphtha boiling point does not go down, as top reflux rate is increased.
3. The crude tower overall delta P increases exponentially, as the crude rate is increased.
4. The diesel product will...?

Pardon me, Professor Pot! I may have a very tiny disagreement! I've inspected the fouling pattern in several crude towers! They're all quite similar.

Here's what I've seen:

- The top one or two trays are pretty dirty.
- The next few trays down are so salted up that the photons from a flashlight cannot pass through the bubbling area.
- Once I crawl below the upper half dozen trays, the tray decks are not all that dirty. In the diesel section, the trays are perfectly clean.

- I guess, Professor, maybe that explains why, from what I see, that when a crude tower floods, the overall tower delta P, does not go up all that much.

What? Certainly fouling rates must be higher in areas of the tower exposed to the higher temperatures? Really, Kumar, rates of coke deposition will double for each increase of 25°K! Elementary!

But Professor Pot! Fouling in crude towers is caused by salting-out and refluxing corrosion products down the tower from the overhead condensers! Actually, the big problem we have in Louisiana refineries is MEA. It's being added to crude as an H_2S scavenger in fracked crudes transported by rail cars from North Dakota! It's the amine salts that are the big problem. Certainly not coke!

HA! Don't be silly, Kumar! Nobody transports crude by rail! It's all pipelined! This is America in the 21st century!

MINIMIZING FLASH ZONE PRESSURE

To maximize diesel recovery from crude, Kumar, one must always maximize two variables:

1.   Heater outlet temperature
2.   Bottoms stripping steam

Kindly take note of these rules!

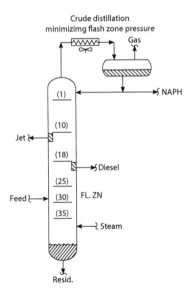

Crude distillation
minimizimg flash zone pressure

Thank you, Sir! I will instruct the unit operators to adhere to these rules! But suppose they ask me if raising the bottoms stripping steam rate, or the heater outlet temperature, causes the flash zone pressure to increase? How should I respond, Professor Pot?

Tell the operators to follow instructions, and not to debate my rules!

I certainly shall, Sir! But I've seen that if one is limited by overhead condenser capacity, that increasing bottoms stripping steam will pressure up the flash zone and suppress diesel yields. Also, as Clare has explained to me, "No sense putting more heat into the tower, if you can't take it out!"

 Clare? I've told you that Indian engineers should not interact with females! It just confuses application of standard engineering principles! Why would more heat and more steam increase the flash zone pressure anyway? Those variables are not related on my crude tower Aspen computer simulation. Go back and check your modeling variables, Kumar!

Professor! What you say is 100% correct! The crude tower pressure should stay constant! The difficulty is that we are limited in the summer by overhead condenser heat removal capacity in our aerial fans!

Nonsense, Kumar! Those overhead air coolers have plenty of spare capacity! You must check their data sheet! Use your head, my dear boy!

Yes, Sir! I shall! But the problems are:

- The air fan belts are slipping!
- The fins are dirty!
- The blade pitch is 10°, not the optimum of 22°!
- The diesel P/A pump is running below its curve value!
- The upper trays in the tower are salted-up!
- The off-gas compressor rotor is also fouled!
- Ambient air is above the design air cooler temperature!
- Steam leaks are blowing into the air cooler fans!
- The ammonia chloride salts are...!

Stop Kumar! I've heard enough! If you would pay more attention to computer modeling of the crude unit, and less to these mundane problems, you would make a better manager! Hmm? Perhaps an email to Dr. Petry would be in order? We should let the Director of Maintenance know about your concerns!

HOW TO ADJUST BOTTOMS STRIPPING STEAM RATE

 Mr. Kumar! The stripping steam in the bottom of a crude distillation unit should always be maximized, up until the point of flooding of the stripper trays or of the upper trays in the crude tower! This is done to maximize diesel recovery from vacuum tower feed! It's a rule!

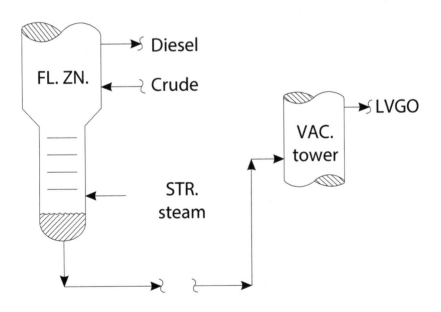

Yes! Thank you for that excellent advice, Professor Pot! But Sir, I have some other concerns in mind. May I mention them?

It's best to follow our rules! And how is your lovely new wife Sonia? Is she a good cook? You're getting a little heavy! Ha! Too much rich curry!

Sonia, unfortunately, does not engage in cooking. She is the lead software engineer for Exxon! But Professor, I've noticed that when I raise the stripping steam rate past 17,000 lbs/hr on my crude unit, some negative factors emerge! Sir, permit me to list my observations:

1. Crude tower pressure increases, due to lack of overhead condenser capacity.
2. The vacuum tower pressure increases, as does the off-gas rate from the seal drum.
3. The LVGO rate goes up.
4. Crude tower diesel production goes down.

Well, my boy! You should make a more vigorous effort to clean your overhead condenser air cooler fin tubes. Make sure the belts are not slipping! Water wash the bottom few rows of tubes, being cautious not to damage the Aluminum fins with too much water pressure!

Check the blade angle – 20° to 25° is best! Try reversing the polarity of the AC motors. That will blow dirt off the lower rows of tubes. Are you remembering to slug water wash the tube side, as we talked about in class?

Yes, Professor Pot! I've been doing all these things you taught us at MIT! The problem is it gets so hot in the summer that often more stripping steam hurts rather than helps. Mainly then, I try to optimize the crude tower bottoms stripping steam rate, to maximize the vacuum in our asphalt vacuum tower!

Yes, Kumar! Rules must be adjusted to fit individual circumstances!

Poor Kumar! I told him not to marry a local girl! Not cooking! Totally unacceptable!

Kumar! Have you considered a water mist to cool the air flow by humidification during hotter summer days? This would reduce the air cooler effluent temperature by about 4°F.

OVERHEAD CONDENSER CORROSION

I see from the maintenance monthly report, Kumar, that you are experiencing an overhead condenser tube failure every few months! You must change to Titanium (Ti) tubes at once! A Titanium tube bundle will not corrode at all!

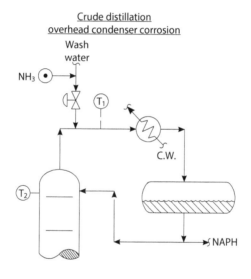

Crude distillation
overhead condenser corrosion

Oh! Professor Pot! I'm afraid that just changing the tube bundle will not stop corrosion in the overhead vapor line, or the condenser shell! Norm recalled that in 1968, at the Amoco Refinery at Texas City, they changed No. 3B pipe still overhead condensers to Ti. Then, the C.S. vapor line failed and almost destroyed the refinery!

Well! Norm should know! He was likely partly responsible for many of the disasters at that terrible refinery. Well, if not Ti – what does Norman suggest?

Of course, the tower top temperature at T2 must be kept 10°F – 20°F above the water dew-pt. temperature, taking into account the partial pressure of ammonium chloride, which may escalate the water dew-pt. temperature by 10°F - 30°F. And then enough water should be added to exceed the rate needed to reach the water forced condensation dew-pt. temperature at T1.

Quite true! Do you recall, Kumar, how to calculate the amount of wash water required to reached the forced condensation water dew-pt. temperature? You can use your Aspen Simulation!

Well, Sir! I have an alternate method! I observe the temperature at T1! Then, I'll increase the wash water rate until the temperature at T1 stops dropping. That, Professor Pot, is the forced condensation dew-point temperature!

Very clever! Of course more water is always helpful! One should design for 2x the wash water rate needed to reach the water dew-pt. temperature at the condenser inlet! By the way, which chemical vendor does Mr. Lieberman recommend, Kumar?

None! He suggests that NH_3 can be used as a neutralizer for the HCl formed from the hydrolysis of the $MgCl_2$ to $Mg(OH)_2$ + HCl! Ammonia is a tiny fraction of the price of neutralizing amines! The NH_3 is mixed with the water wash, before it is sprayed into the vapor line to each condenser.

NH_3! Certainly not! All the chemical vendors recommend a proprietary neutralizing amine!

That Lieberman! Using NH_3! If that worked, why would most refineries use Neutralizing amines? He's a real troublemaker!

ON-LINE SPALLING OF CRUDE PRE-HEAT EXCHANGERS

 I see, Kumar, that your kerosene pumparound vs. cold preheat exchanger has fouled! I've calculated a "U" value of 12 BTU/HR/FT²/°F! Its design "U" is 62. What did you measure the DP on the tube (crude) side at? The tube side calculated DP at the current crude rate is 6 psi!

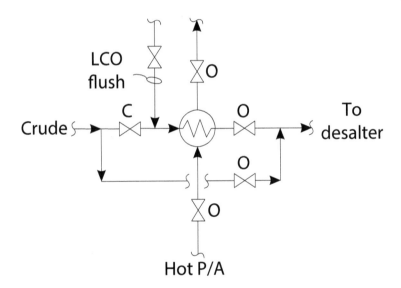

LCO flush

Crude

To desalter

Hot P/A

Professor, the measured DP on the tube crude side was 42 psi! Seven times greater than the calculated DP! This proves, Sir, that you were correct! The exchanger was terribly fouled on its tube side!

I notice, my boy, that you are using the past tense regarding the fouling. Do I also need to instruct you on grammar, in addition to basic engineering theory?

Very sorry, Sir! The exchanger was fouled yesterday, but I cleaned it this morning! Its coefficient is now 38 BTU/HR/FT²/°F, and its pressure drop has decreased from 42 to 16 psi! I should have so informed you! I apologize for my tardy reporting!

What! You cleaned it yourself? Without the maintenance department? John Brundrett, the maintenance supervisor, will be extremely angry! What will I tell him?

Pardon, Professor Pot! Permit me to explain. I employed the following steps to clean the tube side:

1. Block-in the tube inlet, but leave the crude outlet valve open.
2. Leave the hot, shell side flow without change.
3. Open the 1" LCO (or any aromatic type hydrocarbon) flush line. This step is not really all that important.

4. Wait 15 minutes.

5. Return to normal operations!

Norm calls this, "On-line Spalling." It's a method he discovered by accident during a power failure at the Amoco Refinery in Whiting, in 1967. It works great!

And what about the fouling particulates that are "Spalled-off" in this irresponsible, non-standard procedure? Where do they migrate to, if I may ask, Kumar?

Sir! I presume the spalled-off solids, are withdrawn from the desalter vessel during the mud-wash period. I suppose that is why Norm only uses this procedure on crude preheat exchangers, upstream of the desalter?

Hmmm...? I believe, Kumar, that the coefficient of thermal expansion, of the fouling layer, inside the tubes, would be less than the tube metal itself! Thus, the deposits would shear-off, due to the rapid change in temperature, and the sudden restoration of crude oil flow!

I wonder if this technique could be used in other heat exchanger services, such as cracking plants or hydrotreaters? I'll Google it when I get back to the university! After all, the internet is the source of all wisdom and knowledge!

Kumar! May I suggest you have some safety people stand by during this on-line spalling procedure, in case the thermal stresses cause flange leaks to the crude side connections!

EFFECT OF REFLUX ON OVERHEAD ACCUMULATOR TEMPERATURE

Kumar! You have made a small error! The crude unit performance test report shows increasing the top reflux rate, reduced the tower top temperature at T1, but increased the drum temperature at T2? A careless mistake! Of course, the drum temperature must go down, if you have cooled off the top of the tower with more reflux! It's all right, dear boy, the rest of your report was excellent! I have also made errors in my long career! Don't be too upset!

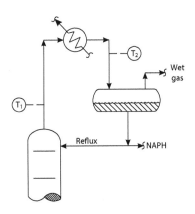

Thank you, kind sir! But it was not an error. I observed this effect in the field, and my observations are in agreement with my heat exchange calculations!

But, but, but? My dear young colleague! Certainly a reduction in the condenser inlet temperature would result in some reduction in the condenser outlet temperature? That is just plain common sense! You must agree?

Perhaps Professor Pot, I am mistaken! But here is my analysis, which God willing, is correct:

- Assume the crude tower flash zone and pumparound duties are constant.
- Therefore the overhead condenser duty (Q) is mostly constant.
- Q = U • A (Delta T)
- Assume (U • A) is constant.
- Therefore, Delta T is mostly constant.
- Also, the condenser's cooling water inlet and outlet temperatures must be constant, since Q is constant.

Quite clever, Kumar! If delta T is constant, then a reduction in the tower top temperature would lead to a reduction in delta T at the hot end of the condenser! Which must result in an increase in the condenser outlet temperature, to increase the cold-end condenser delta T! No wonder you were my prize student at M.I.T. But in practice, have you really observed this hotter reflux drum temperature, due to a greater reflux rate?

Yes, sir! It happens frequently on our crude unit! Also, the hotter drum increases the off-gas rate from the reflux drum, which, unfortunately, is flared! We can stop the flaring at #12 Pipe Still, by increasing the tower top temperature with less reflux. But then, our naphtha end-pt. gets a bit high!

Kumar! Combining theory and observation to draw conclusions is the sign of a brilliant engineer! Well done, my boy!

Such a great misfortune! Kumar should have continued at M.I.T., and earned his Ph.D. I'm afraid his talents are wasted in this midwestern refinery! At least he's married to a wonderful girl from Delhi! I advised his family on the choice of his bride!

It rather seems, Kumar, as if your analysis should also apply as well to the FCU, Visbreaker, and to the Delayed Coker Fractionators!

REMOVING TRAYS FROM PRE-FLASH TOWERS

 Well, Kumar! I see that you wish to remove half of the 12 rectification trays in the crude pre-flash tower! My fine fellow, this is not a good idea! It will negatively affect fractionation efficiency!

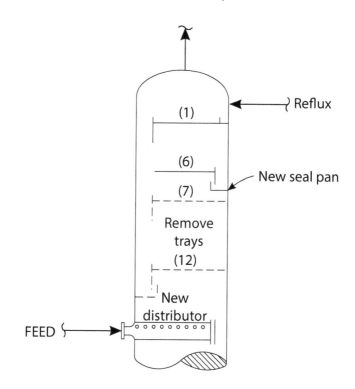

But, Sir! I need to have an additional 12 ft. above the feed inlet nozzle, to allow vertical height, for foam to settle and prevent black naphtha, which ruins the desulfurizer catalyst!

Yes, yes! A worthy objective! Especially crudes that have been delivered by pipeline will have flow-improver chemicals added to them, which may cause such crudes to foam at 350°F-400°F. But Kumar, won't the pre-flash tower naphtha end-point be substantially increased, by removing half the trays?

No, Professor Pot! The extra 6 trays are superfluous! The molal ratio of V/L on these trays is only about 0.3. When the ratio of V/L becomes small, having 12 trays, vs. 6 trays, does not affect the rectification section fractionation efficiency, to any great extent! Sir, it's rather the same concept you taught us at M.I.T., when we studied the McCabe Thiele Diagram. Specifically, when we studied the concepts of "pinch-point" and "minimum reflux rates."

Yes, Kumar! You are correct! Your analysis is also supported by my Aspen Process Model, which shows removing half the trays has only increased the pre-flash naphtha ASTM D-86 95% point by 5°F. Why risk foaming and flooding, for only 5°F worth of fractionation? Good work!

Thank you for your kind words, Professor Pot! My wife has sent some curry for my lunch. May I offer you a portion? It is extremely spicy! Just like we have at home in Mumbai!

Thank you, I'll have just a small cup!

Ah! What a pleasure it is to work with a former student, who applies his Chemical Engineering education to solve refinery process problems! Unfortunately, this curry is far too bland for my taste!

TOTAL TRAP-OUT CHIMNEY TRAY

We must install a new kerosene draw-off in No. 2 Crude Unit during the May turnaround! We can install a draw-off sump below tray #14 for this purpose. Kumar, check that the delta P of the vapor flowing through the valve caps is not less than 2" of liquid, to make sure the tray will not leak and thus allow the by-pass of kerosene around the new draw-off pan sump!

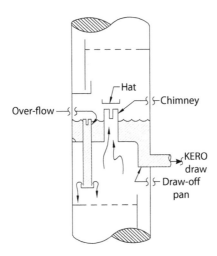

Sir! If this was a five foot diameter crude tower, I would agree with you! But this is a twenty foot diameter tower. The larger the tower diameter, the greater the probability that the tray decks will be out-of-level! If the tray has a sag of several inches, then the tray deck will leak!

Ah, yes! I see your point! We may lose a large quantity of kerosene through the low point of the tray! But that's life, my young friend. You will just have to inspect the trays carefully for levelness during the turnaround!

I am honored, Professor, that you consider me a friend! But, sir, there is also the option of installing a Total Trap-Out Chimney Tray, with provision for internal overflow. Such a tray should have been:

- Sloped, 1"/foot, in direction of flow.
- Sealed to the tray ring, with a 3" x 3" expansion ring.

- Constructed out of 8 gauge, 316 (L) stainless steel, or 317, for additional naphthenic acid protection.
- Water tested for leaks prior to start-up.
- No drain holes.

But Kumar! We will lose a fractionation tray! And also degrade fractionation between kerosene and diesel!

Professor, as usual, you are correct! However, the loss of a single fractionation stage will only increase the D-86 ASTM distillation 5%-95% overlap, from 25°F to 30°F! A small price to pay, to insure the ability of being able to always maximize kerosene production! May I show you my Aspen simulation results?

Agreed! It will be a chimney tray, for the new kerosene product draw-off! I especially like the ideas of sloping the tray in the direction of flow and also the 3" x 3" expansion ring.

If only I can prevent Kumar from being promoted next year into the Planning Division, my life would be complete! The Planning Department will destroy his intellect for certain!

SIDE DRAW-OFF LIMITATIONS

Let us increase diesel production! Kumar, the diesel side draw-off control valve is only open 15%! Open it to 50%! This ought to double the flow of diesel. We are overflowing the top of the chimney! That is very bad! The diesel is running down the tower into the gas oil draw! That's $20/barrel product downgrading, my boy!

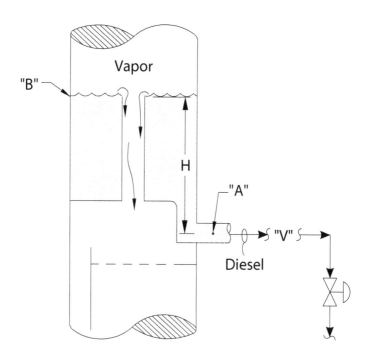

Thank you, Professor Pot, for your wise advice! Most unfortunately, Sir, opening the diesel draw valve will not increase product flow whatsoever! My apologies!

 Well, it's not your fault, dear student! But certainly, as the sump is full, as indicated by the level transmitter, more flow must be possible! Can you not see the truth of my logic?

So sorry, Sir! While your logic is, as usual, brilliant, there is a problem! You see, Professor, that the liquid at point B is in equilibrium with the vapor! The liquid at point B is at its boiling, or bubble point! As this liquid flows through the draw-off nozzle, it accelerates to velocity "V." The energy to accelerate this liquid comes from the head of liquid, H, in accordance with this formula:

$$H = 0.178 \cdot V^2$$

If the pressure at point A, falls below the pressure at point B, the liquid at point A will boil. The giant vapor volume expansion in the nozzle causes vapor lock, and prevents any increase of flow!

 Brilliant analysis! But about the 0.178 coefficient! How was this derived? Let us assume H is expressed in inches, and V is expressed in ft/sec!

Well Professor Pot, it's a theoretical coefficient, based on converting potential energy into kinetic energy! It makes no allowance for friction in the draw-off nozzle! Norm says he uses a 0.30 coefficient in sizing new nozzles, to allow for friction! But this assumes that there is no vortex breaker in the nozzle!

 But of course, Kumar! If the nozzle exit velocity exceeds 4 ft/sec, a vortex breaker is most certainly required! However, for velocities of less than 2 ft/sec, vortex breakers are best omitted!

  How I do wish that Kumar would not associate with that Norm! He's a bad influence on young people! He has no respect for authority or scholarship!

Kumar! Will you join me in a cup of green tea? I believe that the problem you have described is referred to as "Nozzle Exit Loss Cavitation Limit"!

PACKED TOWERS

Kumar! We must always strive to use the best available technology in our designs! For that reason, we should prepare a revamp design using beds of structured packing and high-capacity perforated rings for our crude distillation column!

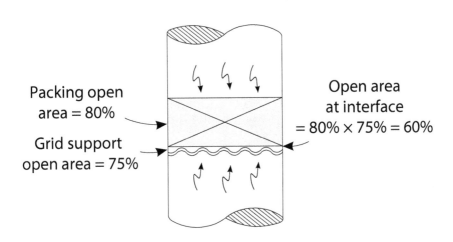

Packing open area = 80%

Grid support open area = 75%

Open area at interface = 80% × 75% = 60%

Thank you, Professor, for your wise advice! But sir, for what purpose do you suggest we remove the existing grid trays, and replace them with beds of packing?

Dear student! As usual, you have asked a perceptive question! The rings or dumped type packing will be used in the fractionation zones. Dumped type packing may have an HEPT (height equivalent of a theoretical plate) of as little as 20 inches! Likely, this would give us double the number of fractionation stages for the same height of trays! The structured packing would be for the pumparound section, where it would increase capacity, compared to multi-pass trays, by 30%!

Yes, Professor Pot! Packing of various types are in widespread use in modern process plants throughout both India and America! But I've asked Norm about his experiences! He has graciously bestowed on me his ten "Rules of Thumb for Packed Towers." May I be permitted to email you Norm's rules?

RULES OF THUMB FOR PACKED TOWERS

1. Packing capacity is proportional to the open area of the packing, times the open area of the grid support. NOT, just the open area of the packing alone.

2. Fouling occurs at interface between grid support and packing. And, at the top of the packing.

3. Dumped packing cannot be inspected while packing is being loaded. Don't use rings or saddles.

4. Each layer of structured packing must be inspected after it is laid down. Watch that it is not crushed while the upper layer is installed.

5. Always put a layer or two of grid between the packing support and the packing.

6. Packing does NOT redistribute liquid. Small liquid rates (a few gpm/ft^2) are very tricky to distribute.

7. Always use a hold-down grid, above packing, but keep it a few inches below liquid distributor.

8. Do not use orifice plate, chimney tray type liquid distributors. They lead to bad liquid distribution if slightly out of level.

9. Provide for internal overflow for total trap-out chimney tray - vapor distributors. Do not overflow through chimney slots.

10. Don't use packed towers. The final installation cannot be inspected. The best choice for new towers is ancient BUBBLE CAPS, which out-perform in efficiency (but not capacity) all new tray developments.

Ha-ha! Bubble cap trays! Norm should always be treated with respect due to his advanced age! We, as Indians, must always respect our elders. But the use of bubble cap trays in crude distillation was already outmoded technology when Norm began his career in 1965 for the American Oil Company, in Whiting, Indiana!

I must really ask Lieberman to desist from his efforts to influence Kumar with his archaic, outmoded engineering concepts such as the use of ancient bubble cap trays!

Author's Note – I last ran into bubble cap trays at a refinery in New Jersey, built in 1954. They were welded-in trays, could not be replaced, and appear to be working reasonably well after over 60 years of service.

CONTROLLING AGO DRAW-OFF RATE

Kumar – dear friend, colleague, and former student. Kindly obtain a sample of atmospheric gas oil from the lower side-cut of our crude distillation tower! Submit it to the laboratory for Ni and Va! Also for conradson carbon. The FCU is complaining, once again, as to the quality of their feed! They suspect it's our AGO that is the culprit that is degrading their catalyst activity!

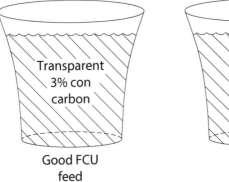
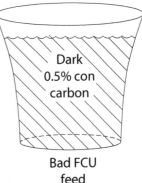

Transparent
3% con
carbon

Dark
0.5% con
carbon

Good FCU
feed

Bad FCU
feed

No need for a lab sample, Professor! Our AGO sample is transparent! This indicates that it is free of asphaltenes. It is these highly condensed aromatic rings that make up the asphaltenes that contain the Ni and Va, that contaminate FCU catalyst! They're part of the molecular structure of the asphaltenes!

But Kumar, what you say is not correct! I myself, with my own eyes, have just seen in the Shift Foreman's office a bottle of recently drawn AGO! The sample was quite greenish, and totally opaque! Certainly not transparent as you have described!

So sorry to be argumentative, sir! You cannot judge the quality of gas oils – either AGO, LVGO, or HVGO, by their appearance, when they are cold, in a large container, such as a pint sample bottle. The sample must be obtained in a small, 1 oz. bottle, and heated to the temperature of hot jasmine tea, so as to melt the wax crystals! As long as such a sample is transparent, it is free of asphaltenes! Hence, also free of metals! Even 4 con carbon HVGO will be excellent FCU feed, if it is transparent, as I have described! Norm produced such FCU feed in Aruba, for a FCU feed hydro treater, and it was fine!

Here in America, we must say we heat the gas oil sample in a cup of hot coffee – not tea – as we would do in India! We must always accommodate ourselves to local customs! Ha ha ha!

Hmm? But all our vendor catalyst guarantees are based on a maximum 0.3 con carbon. This is inconsistent with what Kumar has heard from Norm. Most likely, Norm is confused. Unless the conradson carbon content of gas oil is volatile?

Kumar! Perhaps we had best base our gas oil quality on percent hexane insolubles, as a better measure of asphaltene content than concarbon!

FILMING AMINE PLUGS OVERHEAD VAPOR LINE

Kumar! Your recent crude tower overhead pressure drop survey shows a 4 psi. That is, DP = 25 psig minus 21 psig! Have you calculated the vapor outlet nozzle exit loss using the correct formula?

$$\bullet \, DP\,(PSI) = 0.3 \times \frac{Density\ Vapor}{62.3} \times \frac{(V)^2}{27.2}$$

Where V = Nozzle Velocity, in ft. per second

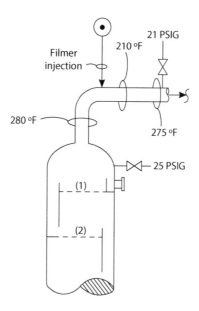

Yes, Professor, I have! The calculated nozzle exit loss is one psi!

No Kumar! Very sorry! The observed and calculated nozzle DP should be the same! I believe, my boy, that you have made an error – either in your calculations, or in your measurements.

Pardon me, Sir, if I may take the liberty to observe, that the 16" crude tower overhead vapor line is partially plugged! Apparently, the filming amine is restricting the vapor flow and causing an excessive pressure drop!

How can you know that, Kumar? Have you developed X-ray vision? Can you now see through steel pipes – like Lois Lane? She is Superman's...!

Pardon, Professor! Miss Lane is Superman's girlfriend. However, their relationship is purely platonic! My observation that the vapor line was partly plugged, downstream of the filming amine injection point, was based on a skin temperature survey! The localized cold section of 210°F – about two or three feet long, could only be attributed to a local fouling layer! This is a common problem on very many crude units that employ filming amine!

Hmm? This is bad! A high vapor line pressure drop will increase the crude tower flash zone pressure, and reduce the diesel recovery from the vacuum tower feed! That extra 3 psi DP may be costing us $30,000 a day! We should ask the Operations Division to cease the use of filming amine as soon as possible. I shall email Mr. Frank Citek at once. Only yesterday Mr. Citek criticized me for the lack of tech service contribution to refinery profitability!

But Sir! The filming amine is quite useful in protecting the overhead vapor line from corrosion, due to the HCl evolved from the hydrolysis of $Mg\,(Cl)_2$ salts in crude! It is not so much that the filming amine is at fault, but that the crude unit operators are not consistently adding the dispersion naphtha into the flow of the filming amine! Perhaps a note to Mr. Citek, pertaining to the importance of the dispersion naphtha would be useful, as he is the operation's manager.

 Yes! I quite agree! Kindly draft such an email to Frank Citek. But make sure you explain the difference between filming amine and water soluble neutralizing amine! I fear Mr. Citek does not know the difference!

Note from Norm

Frank Citek was Amoco East Plant Operation's manager in Texas City in the 1970s. I remember him screaming at me one day, "Lieberman! Stop complaining! We make our own luck!" The other lesson he taught me was, "The performance you get from people is not what you hope or expect, but the minimum you are willing to tolerate!"

STRIPPING TRAY PRESSURE DROP PROFILE

Dear Kumar! Your pressure drop survey on the crude unit distillation tower bottom stripping section is in error! My boy, the pressure you have shown above tray #4, is one psi larger than below tray #1! This is quite impossible! Please revisit your field measurements! You may have accidentally reversed your pressures.

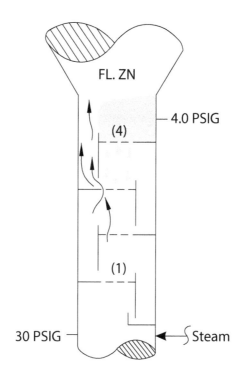

Sorry Sir! I was quite careful! The pressure above tray #4 is slightly higher than below tray #1.

Very well! But how can you explain your results? Certainly, vapor cannot flow on its own, from 3 to 4 psig?

Certainly Sir, you are correct! My explanation is that we have lost the downcomer seal from tray #4. Likely the downcomer is too short, or the outlet weir from tray #3 is badly out-of-level!

Yes, Kumar! These sorts of installation problems are quite common. Or the tray #4 tray deck might be damaged? Who knows? Tray installations in this country are often done carelessly. Not like in India!

Yes, Professor! And then, the liquid would stack up above tray #4, because it would have to flow through the sieve tray's $\frac{1}{2}$" orifice holes or the grid tray's fixed valve caps! If the s.g. of the liquid was 0.70, then a back-up of 40 inches of liquid above tray #4 would indeed produce a head pressure of one psi!

Ah yes, dear student! You can confirm your theory by measuring the pressure in the flash zone! It should read a bit less than 3.0 psig! Also, I imagine that the stripping efficiency of bottoms is poor?

You are correct Sir, on both counts! I checked the ASTM D-86 distillation of the crude tower residue. It's 16% 650°F and lighter! That is diesel, kerosene, and even naphtha is being lost to the vacuum tower feed! And then these valuable distillates flow straight on to the FCU with the LVGO product! Terrible!

Disgraceful! This could never happen in Mumbai! But once Donald Trump becomes President of the United States, such inefficiencies will be eliminated!

Hmm...? I wonder if I should apply for an extension of my visa...! It's somewhat outdated! I don't wish to be deported by the new President Trump!

PROTECTING CRUDE TOWER STRIPPING TRAYS FROM DAMAGE DUE TO WATER IN STEAM

Kumar! Bad news! The Inspection Department has reported that the crude tower bottom stripping trays have been found dislodged from their tray ring supports once again! Dr. Petry is quite concerned!

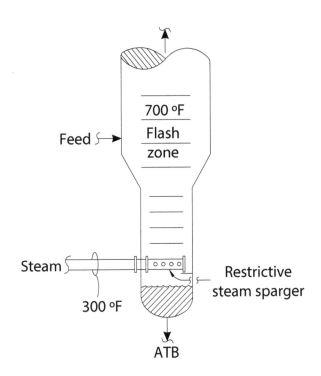

Professor! We need to install a restrictive steam sparger!

A what? Is that like a steam distributor – or is it something else? What is its purpose, Kumar?

Well yes, Professor Pot! It's rather like a steam distributor. However, the total hole open area is quite small! Hole velocities, for a crude tower, are approximately 200 to 400 plus feet per second! The hole pressure drop is a few psi. The distributor pipe is made of "triple X" thick pipe! Then, in case a slug of water enters the sparger, along with the stripping steam, the sparger will pressure-up, as the water flashes, and trips off the supply of steam to the stripper!

So, a new trip valve is required for this concept? That would be quite costly, my boy!

No, Sir! We would simply use the existing steam flow FRC as the trip valve! Professor, you can see the details of the design, which has been used at many refineries, in the Appendix to the 2nd Edition of *Process Design for Reliable Operations*.

And who, dear student, is the author of this text? It's not that Lieberman fellow, I hope?

Unfortunately, Sir, this book was authored by Mr. Lieberman! Still, his design has been used with good results in vacuum towers in California, Louisiana, and Aruba! Also, Professor, Norm says we should use for the stripping tray design for improved mechanical integrity:

- Explosion doors in tray decks
- 10 gauge trays (4 mm) thickness
- Shear clips
- Back-to-back tray decks
- Bolted-in tray panels
- Grid tray, rather than valve or sieve tray decks

 Feet per second? PSI? Why not meters and bars? These Americans, with their BTU's and °F and pounds. They should change to SSI units like the civilized world. They're so primitive! But, I suppose in the long run, they will pay for their retrogrid units themselves!

Fine, Kumar! But I imagine this will double both the cost and installation time of these exotic replacement trays!

CHAPTER THREE

ENGINEERING BASICS

IRENE & NORM

Good Morning! I'm Irene, Norm's very smart daughter!

Hi! I'm Norm, Irene's dad! I've taught Irene everything I know!

DRAFT IN FIRED HEATERS

 Dad! What's draft, anyway?

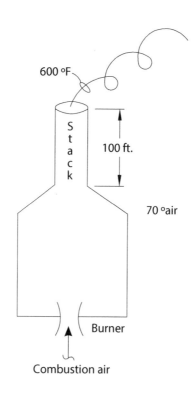

It's the pressure difference between the firebox and the atmosphere, measured at the same elevation. That means that the pressure inside the furnace is below the pressure in the atmosphere. Hence, combustion air is sucked into the firebox through the burners!

Hmm? ... OK, Dad! But what causes that draft? Our professor at LSU said draft is caused by the tendency of a heater to "suck-in" air. But what does that mean?

That's not true! Draft is due to the temperature difference between the 600°F flue gas, and the 60°F ambient air!

Temperature difference? Dad, does this have anything to do with what I learned in grade school – "Hot Air Rises"?

Yes, Irene! Hot air rises because it's lighter! It's less dense! What causes draft are two things! One of these things is the height of the stack. The taller the stack, the more the draft! The other thing is the density difference between the colder, outside air, and the hotter flue gas in the stack!

Dad. Bobby at school said you're full of hot air! Ha-ha, ha-ha! I guess what you're saying is that if I get the flue gas hotter, the draft will go up and the heater can suck in combustion air! Is that right?

You are correct, Irene! If you double the height of the stack, from 100 ft. to 200 ft., you'll double the draft! And if you increase the flue gas temperature from 600°F to 1,200°F, you will …

Dad, Dad! I know! I know! You'll double the draft! I'm so smart!

Sorry, Irene! You have to first convert from °F to °R (Rankine). Let me write it down for you…again!

- Ambient air at 70°F + 460 = 530°R
- Flue gas at 600°F + 460 = 1,060°R
- Draft is proportional to: 1,060/530 = 2.0
- Flue gas at 1,200°F + 460 = 1,660°R
- Draft proportional to 1,600/530 = 3.14
- 3.14 – 2.00/2.00 = 57%

Irene, you can see that the draft did not double when you heated the flue gas from 600°F to 1,200°F, but only went up by 57%! You forgot about °R. If we lived in France, you would use °K (273°K = 0°C), instead of °R (460°R = 0°F)! Kelvin instead of Rankine!

Dad! I'm very interested in this! Really! Could you loan me $2,800? I'd love to go to Paris to study Kelvin! I've always had a secret desire to study the metric system!

My dad is such a nerd! I mean, everyone knows the wind also makes a lot of draft – but he didn't even mention that, 'cause he didn't read it in some dumb book!

Technical Follow-Up to Calculate Draft:

1. Calculate density of air outside the stack, lbs/ft^3.
2. Calculate density of flue gas inside the stack. Assume M.W. of flue gas = air, lbs/ft^3.
3. Take density difference between cold air minus hot flue gas, lbs/ft^3.
4. Multiply density difference by height of stack, ft.
5. To convert to inches of water draft:
   a. Divide by 144 sq. inches per sq. foot to get psi.
   b. Multiply by 27.7 inches of water per psi.

Note, that the effect of frictional losses in the stack and the effect of wind, have been neglected in the above calculations.

TURBINE EXHAUST SURFACE CONDENSER OUTLET TEMPERATURE

 Dad! A steam turbine exhaust is condensed in a gigantic surface condenser! The colder the surface condenser outlet, the better! Is that right, Dad? Colder is best?

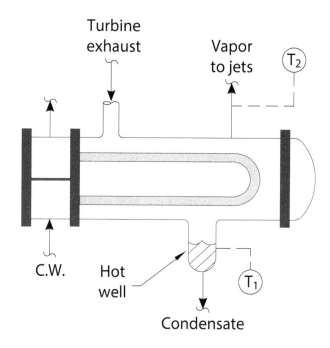

Yes, Irene! The colder the condenser, the lower the pressure of the condensing steam! You can extract maybe 10% more work from the turbine's steam if you lower the surface condenser outlet temperature by around 20°F - 30°F!

Gee. Dad! That's a lot of energy! But I've got another question. Which outlet are you talking about? The vapor outlet (T_2) going to the ejectors or the water temperature (T_1), in the hot well boot?

Irene! Really! Isn't it obvious that it's the vapor outlet temperature that controls the **condensing pressure** of the turbine exhaust steam! There are all sorts of problems that can cause the water in the hot well (boot) to be **SUB-COOLED**. But sub-cooling the steam condensate will **not** reduce the surface condenser pressure! Will it?

Hmm?... But, Dad? Every book I've ever seen, and every surface condenser in the entire, great big world, has the temperature indication in the hot well at T_1, and never, ever, in the vapor outlet line at T_2. It's backwards from what you say.

Every design in the world is wrong! Only what I say is right! The temperature indication is always located in the hot well for historical reasons! It's a long-standing design error!

Okay, Dad! I suppose if **YOU** say it, it's right ... because it's you who has said it, and you know everything!

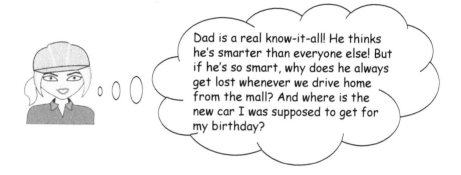

Dad is a real know-it-all! He thinks he's smarter than everyone else! But if he's so smart, why does he always get lost whenever we drive home from the mall? And where is the new car I was supposed to get for my birthday?

Technical Follow-Up

To calculate the effect of surface condenser vapor outlet temperature (T_2), on horsepower extracted from turbine steam:

1. Obtain Mollier Diagram.
2. At motive steam pressure and temperature, read vertical axis, BTU per lb.
3. Drop straight down on line of constant entropy until intersect line for condenser vapor outlet temperature or pressure. If confused, just check vapor pressure of water at this temperature on steam table. Use this as pressure in surface condenser. Read vertical axis, BTU per lb.

4. Divide BTU difference between steps 2 and 3
    above by 2,457 BTU per horsepower. That's the
    horsepower you extract from each pound of steam
    to the turbine, at 100% efficiency. The reciprocal
    (i.e., pounds of steam per horsepower) is called the
    "Water Rate".

ADJUSTING STEAM TURBINE TO SAVE STEAM

 Dad! I heard Carl and Aunt Clare talking about adjusting steam turbine speed! According to Aunt Clare, it's best to slow down the turbine governor speed control valve to keep the pump's discharge control valve in a mostly open, but still controllable position!

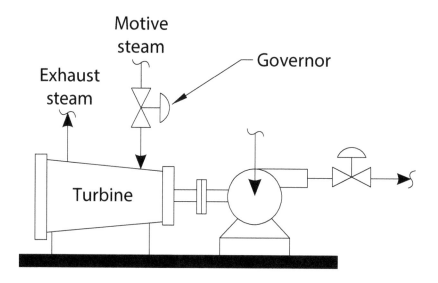

Yes, Irene! That's very smart of you!

Yes, it's true! I'm very, very smart! But, Dad? How do you figure out how much steam you'll save by slowing down the turbine? Professor Pot says you have to use the **AFFINITY LAW**! What's that?

The affinity law states that work is proportional to the rotating speed of the turbine, raised to the 3rd power! Cubed! For example, if I slowed the turbine down from 3,300 rpm to 3,000 rpm, I would save ...

Dad! Dad! I know ... 10%! See how fast I got the answer? Ha-ha!

No, Irene! It's like this:

100% - [3,300 – 3,000/3,300]³ x 100% = 25%

Boy, you sure do make things super complicated! I guess that's why you have no friends? So I guess that will cause the governor speed steam control valve to close? But that's also bad, I think. Won't steam energy be wasted, if that governor turbine speed control valve closes some?

That's a big problem! It's not so much that the steam's energy is lost. That remains constant. It's that reducing the steam pressure across the turbine's governor inlet control valve reduces our ability to extract work from each pound of steam! We say that the **ENTROPY** of the steam has been increased! To avoid this loss, we have to close off a **port valve** on the steam chest in order to...

Dad! Dad! Stop! Don't you start with that entropy stuff again! I heard that from Professor Pot last year, and I don't need to hear it again from you! Isn't a port valve what Auntie Clare calls a hand valve, or a horsepower valve?

Technical Follow-Up

To adjust turbine speed to save energy, proceed as follows:

- Step 1 – Slow down the turbine speed until the pump's discharge control valve is about 75% - 80% open.
- Step 2 – Close completely, one of the port valves on the steam turbine chest. Measure the steam chest pressure on the condensate drain line, if there is no gauge on the chest itself.
- Step 3 – Close additional port valves, until the steam chest pressure rises to within 10 to 25 psi of the motive steam pressure in the supply header. You will have two or three port valves to work with. When you close one, shut it tightly. If the governor is wide open, you have closed too many port valves.

  If closing a port valve does not force the governor speed control valve to partly open, it means that the seat of the port valve is destroyed through improper prior use!

DISTILLATION TRAY DOWNCOMER SEAL

Dad! Why is it that trays have weirs? Actually, that's not my question! Why is that tray weirs are like ½" higher than the bottom edge of the downcomer, from the tray above?

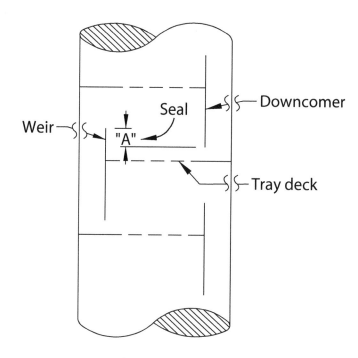

I guess, Irene, you're asking about dimension "A"? That's called the downcomer mechanical **seal**! It's needed to keep the bottom edge of the downcomer submerged in the liquid on the tray below! The purpose of this seal is to keep the vapors from blowing up the downcomer!

Hmm...? I guess I got that okay. But, Dad! Without using any big words, could you just explain why that would be so bad, if the vapor bubbled-up through the downcomer?

For two reasons! The first is that vapor would by-pass the tray above! The vapor would then not efficiently mix with the refluxed liquid on the tray! The second reason is that the vapor would push the liquid back-up, out of the downcomer, and flood the tray above! Understand, Irene?

Not really! What's the big deal if one tray floods? There's a whole lot of other trays in the column. They'll just have to work a little bit harder! Like when I miss a day of school, I just have to work a little bit harder to catch up!

No, Irene! If one tray floods due to downcomer back-up, because of a loss of its downcomer seal, then two bad things are going to happen:

1. All the trays above, with time, will also flood!
2. All the trays below, will dry out!

And dry trays and flooded trays do not fractionate! Understand, Irene?

I suppose getting dimension "A" too small would also cause too much downcomer back-up and flooding? But, Dad! Why, when I ask you a question, does your answer always have to be in numbered parts?

Dad is so strange! He told me last week that he even counts his steps when he goes running – 6,668 steps make his regular four-mile run! I wonder if his other engineering friends are just as nerdy? He actually thinks counting steps is more accurate than my GPS watch!

BERNOULLI'S EQUATION

Irene! Are you learning anything at university, or is this all a complete waste of time and money? $40,000 a year tuition! That's big bucks!

No friction

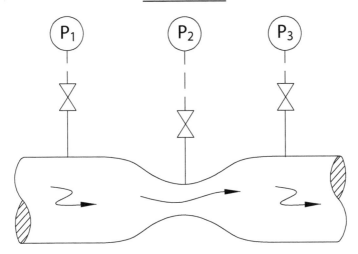

$P_1 = 100 \text{ PSIG}$
$P_2 = ?$
$P_3 = ?$

 Sure, I'm learning lots of good stuff! Like about Daniel Bernoulli! He was born on February 8 – but I forgot what year. He worked in Switzerland. His equation is all about hydraulics – I think?

Okay, look at my sketch! If I've got water flowing through a pipe at a constant rate, and the pipe is really smooth, so that there is zero friction. And the pressure at P1 is 100 psig, what's the pressure at P2? Higher than 100 psig, just 100 psig, or lower? Remember, Irene, that there are no frictional losses at all!

Well, Dad. No friction? I guess the pressure at P2 would have to be exactly the same as P1 – 100 psig! Where's my prize for a right answer?

Irene! How about acceleration? At P1, the water is moving at six ft/sec! When the water flows from a four-inch pipe into a one-inch pipe, it accelerates to 96 feet per second. That energy, according to Dan Bernoulli, comes from the pressure of the water:

- Delta H is proportional to (velocity)^2
  or
- Delta H = 0.178 × (V)^2

  Where

- Delta H = potential energy, inches of liquid head change

- V = Velocity in feet per second
- 0.178 = theoretical orifice coefficient. More typically we use 0.3 to 0.6, to allow for some friction and turbulence.

Hmm? I'm thinking! So, I guess the pressure at P2 would be lots lower than the pressure at P1! Maybe like 70 psig! And since I'm really very, very smart, I'm thinking the pressure at P3 would be exactly the pressure at P1! 100 psig! Right?

Yes, Irene! That's called pressure recovery. It's 100 psig, because there is no friction!

Seems like I learned this stuff in high school for free! Now it's $40,000? I guess that's the price we pay in America for cell phones, computers, emails, and video games!

Now see if you can't calculate the theoretical coefficient for conversion of kinetic energy, to potential energy of 0.178, using the acceleration due to gravity on our home planet! (32.2 ft/sec/sec).

PROPERTIES OF STEAM

Dad! Why is steam so important? That's all
I hear about in engineering school! Steam!
Steam! Steam!

Converting heat to velocity

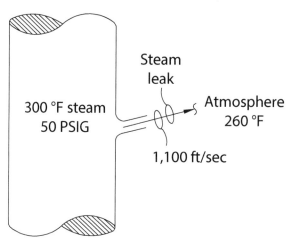

That's because we, as humans on the planet
earth, use steam and water as our working
fluid! Also, a pound of steam has a lot more
recoverable energy than any other material!
Not because of its sensible heat, but because
of its latent heat! To cool one pound of steam at
300°F, to one pound of 200°F water releases:

- Sensible heat = 55 BTU
- Latent heat = 940 BTU

To cool one pound of gasoline vapor from 300°F to 200°F of liquid gasoline releases:

- Sensible heat = 60 BTU
- Latent heat = 130 BTU

So, Dad, I was wondering. When steam blows out of a hole really fast, I guess that the energy to make the steam move so fast – I guess to accelerate the steam, that energy must come from the pressure of the steam? Kind of like what you explained before. Converting pressure into kinetic energy?

No! Not correct! The energy to accelerate the steam comes almost entirely, like about 98%, from the heat content – or enthalpy – of the steam. That's why a steam leak feels relatively cool, compared to the steam's temperature inside of the pipe!

I see! And since steam has so much heat – in the form of latent heat – that's why it can speed up so much! I bet that as steam blows out of a hole in a pipe, it even partly condenses! That must be why steam leaks have water in them, even though the steam itself in the pipe is dry! I think, Dad, that I must be getting really smart?

Yes, Irene! You're the second smartest of my two daughters! And when you add water to saturated steam at a constant pressure, what happens to the steam-water mixture? Does it get colder or hotter?

Dad! Dad! I know! The mixture temperature remains the same, because steam is a pure component! Water boils and condenses at the same temperature, as long as the pressure is the same! You would have to get rid of all the latent heat first, before the mixture could be cooled! I'm so very, very smart!

MEASURING FLOWS

Why, Dad, do we have orifice plates and flanges? If the orifice plates are flow indicators, how does this work? How can a plate with a 0.45" hole in it know what the flow is?

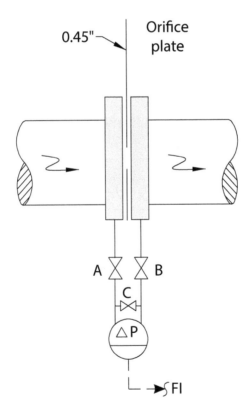

We do not measure flows at all! We measure delta P's! The pressure drop is measured between the two orifice flanges! Then, we apply the formula:

$$\text{Delta P} = K \cdot V^2 \tag{1}$$

A typical orifice coefficient (K) is about 0.5-0.6. The Delta P is measured by the delta P indicator, which has been calibrated for some presumed density, in inches of liquid! "V", is the velocity in feet per second!

Then we solve for "V" in equation (1). Next, on the upstream side of the orifice plate handle there's a number like — 0.48" —. That's the size of the hole in the orifice plate! And then...!

> And then I'll multiply the velocity by the area of the hole:

$$\frac{V \times (0.48)^2 \times \pi/4}{144 \text{ sq. inches per sq. ft.}} = \text{Flow (ft}^3\text{/sec)}$$

> Wow! I'm so smart, sometimes I amaze myself! But, Dad! I've seen times when the hole size on the orifice plate handle was on the downstream side! Does that matter any? It's still just the same sized hole!

Yes, that's important! The beveled or rounded edge of the orifice plate is supposed to be on the downstream side! If the plate is reversed, DP will increase by around 20%, and the observed flow will be around 10% too high!

Also Irene, check that the meter is not off-zero! Close valves "A" and "B". Then open "C"! There's a screw on the DP meter that you can turn to zero-out the flow! Also, in case the taps plug, you can...!

 STOP! Why, Dad, do you have to make everything so complicated? I guess that's why you have such few friends?

 Actually, Dad doesn't have any friends – unless you count his clients as friends. If he had any friends, he would send them an invoice just for talking to him!

HEAD LOSS IN PIPELINES

Dad! Today at LSU we learned from Professor Albright about head loss in pipelines! It was very, very complicated! I have absolutely no idea what he was saying! Dad, do you understand it? Whose "HEAD" is being lost anyway?

Kevin the horse
Work at rate of
one hours power
or
2457 BTU/HR
or
33,000 ft-LBS/minute

Forget about what that professor said! It's really very simple. Let's say I dig a ditch from our house, down to the swamp! Our house is ten feet above the water level in the swamp, which is about a mile from our house. Now, I run water from our house down to the swamp, through the ditch, at a rate of 10 GPM! You and I stand looking at the water running down the ditch from our house to the swamp and you say...-

I know! I know! We've lost ten feet of head in the mile of ditch at a flow rate of 10 GPM of water! But, Dad, we're talking about pipes! Not your dumb ditch! And, where is that pressure going to anyway? I thought you're always saying that energy is conserved? The more I think about it, the more confused I get! Help!

Calm down, Irene! If I replace the ditch with a pipe, I would simply say that I have lost ten feet of head, or elevation, per mile of pipeline! The ten feet of head is potential energy. The head lost due to friction, goes to heat! So, both in the pipeline and in the ditch, we have converted potential energy to friction! If I had one pound of water lose 772 feet of elevation in a pipeline due to frictional loss, the water would heat up by 1°F. That 772 ft-lbs of potential energy, is equivalent to one BTU of heat, was discovered by Professor Joule in England and also Dr. Mayer in Bavaria. Professor Joule determined the equivalence of heat and work by experimentation. Dr. Mayer, more cleverly, calculated it from Cp-Cv, the ratio of specific heats being proportional to—

Stop, Dad! Stop! You're really just as bad as Professor Albright! Can't you just answer the question without showing off?

 Actually, Dad's worse than Professor Albright! I wonder if Dr. Joule had any kids who used to worry about head loss in pipelines!

IRENE EXPLAINS HORSEPOWER

(With the help of Kevin the horse)

- Horsepower is a rate of expending energy to do work.
- Work means lifting a weight of one pound in one hour by "X" feet.
- If x = 2 million feet, then our horse, Kevin, will be doing one horsepower of work, if our horse does the lift, in one single hour.
- More practical, if our horse is lifting a ton of straw up to a hay loft 500 ft. high, then the horse is doing a single horsepower of work, if he takes one hour for the lift.
- Or, if the horse lifts the ton of hay 50 ft. in six minutes.
- Thank you Kevin for your help.

REFRIGERATION

 Dad! I was just wondering, how does our home air conditioner work? They said in school it's just like a refrigeration cycle.

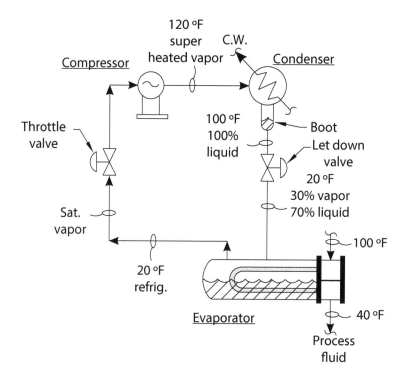

Look at the sketch! You have an:
- Evaporator
- Condenser
- Compressor
- Let Down Valve

Very interesting! Very, very confusing! What's the let down valve for? What happens if it opens too much?

Then you'll blow uncondensed refrigerant right back into the suction of the compressor and waste compression energy and capacity!

Yeah! The refrigerant vapors spin around to no purpose! But what happens if the let down valve closes too much? Is that also bad?

Of course! Then the refrigerant backs up into the condenser! The Freon covers tubes, and makes the area of the condenser that can actually condense the compressor discharge flow smaller. The rate of condensation goes down, and the compressor discharge pressure goes up, and the compressor horsepower increases, and the—

Stop! Stop! One idea at a time! No wonder no one understands you! I'll ask you another question! But this time, Dad, keep your answer simple, okay? Here's my question – what controls the level in the condenser's boot?

Refrigerant inventory!

Good job, Dad! Then what controls the process fluid outlet temperature at the 40°F? Be careful! Keep your answer simple!

The compressor suction throttle valve! Then the refrigerant level in the evaporator will be a function of the—

Stop! You're doing it again! Next you'll start telling me about quantum mechanics, string theory, and all that other incomprehensible stuff you think people are interested in, but aren't!

Actually, I did want to ask Dad about optimizing the refrigerant composition, and handling noncondensable build-up in the circulating refrigerant, and where to drain lube oil contamination from the refrigerant loop! But I really can't take anymore of Dad's long, long explanations today!

Hey, Dad! But here's a question you can probably answer! Can I borrow the BMW tonight? Also, I'll need gas money!

PACKED TOWERS VS. TRAYED TOWERS

 We learned in school today that packed towers are better than trayed towers! Isn't that right, Dad? Especially for fractionation towers?

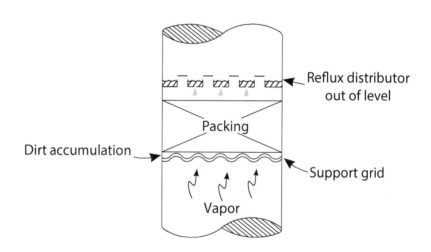

Who told you that bit of stupidity? Packed towers are evil! At least in fractionation service! Not so bad in pumparound heat removal services!

It was Professor Pot! Don't you like him, either? Professor Pot told us in school that packing typically has an HETP – that's an abbreviation for Height Equivalent to a Theoretical Plate, of about 18" to 20". For a trayed tower, he told us we would need like 36" to 48"! Wow! That means I could shrink a distillation tower in half by using packing!

Also, Professor Pot says packing has like 20% to 40% more capacity than trays! So you can make a distillation tower skinnier, as well as shorter! Isn't that terrific?

That's all the junk that Professor Pot learned at MIT – Mumbai Institute of Technology. Here's why I hate packing:

#1 – The packing is supported by a grid. The capacity of the packing is proportional to the open area of the grid, times the open area of the packing – not the open area of just the packing!

#2 – Dirt accumulates between the packing and the support grid. And that creates a pinch-point, where flooding starts!

#3 – Once packing is installed, it can't be inspected for dirt, crushed rings, plastic bags, and discarded scaffold boards!

 Dad! I've asked you before not to number your answers! Anyway, why do you have to be so negative? You ought to look on...!

Irene! Can I finish?

#4 – For packing to work correctly, the reflux must be carefully distributed. Otherwise, the HETP can be far, far larger than for a trayed tower! Don't forget each tray also acts like a vapor-liquid re-distributor! Liquid distributors for packed beds, in fractionation service must be gravity distributors, and not simple spray nozzle type distributors!

#5 – Spray nozzle distributors are only suitable for pumparound services, or wash oil – not for fractionation zones!

#6 – Packing is...!

Okay, okay! I'll never, ever, mention packed towers in our house again! But do you have enough miles in your Frequent Flyer program to get me a business-class ticket, or do I have to fly to Paris economy class with my friends this summer?

I'm so glad that I never told Professor Pot who my father is! I'm sure he would never give me an "A" if he actually knew my dad! If he asks, I'll tell him I'm adopted!

CONDENSING STEAM TURBINE EXHAUST

Dad! We went on a class trip to a fertilizer plant in Oklahoma last week. They have gigantic steam turbines there. The steam turbines all have their exhaust steam outlets going to vacuum surface condensers! What I don't understand is that some of the turbines have their exhaust steam coming off the top of the turbine. Others off the bottom! Which is best? Do you know? You must know, as you claim that you know everything!

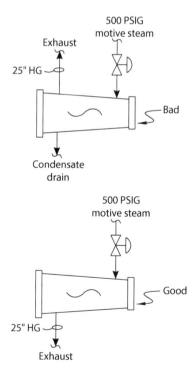

Look at your Mollier Diagram, Irene! If the steam is exhausted to 25" Hg or 125 mm of Hg, and if the turbine is efficient, then the exhaust steam could have 5% to 10% water! Remember your Thermo class – Isentropic expansion! Then if—

But, Dad! If the exhaust is on top, then how could the water formed from the exhaust steam ever get out of the turbine case? I'm confused!

Irene, it's a problem! It's a bad design! The condensate has to drain into a pot located below the turbine! Then the condensate has to be either pumped away or revaporized away with steam! There are all kinds of problems that develop with the condensate drain line – plugging, vapor lock, air leaks ...

Okay, I get it! But what happens if the water can't drain out of the turbine case really completely? That can't be too good for the turbine's rotor? I bet the spinning wheels don't like churning around in water!

The turbine will vibrate! To stop the vibration, the operators will restrict the motive steam flow! Look at your Mollier Diagram! As you close off on the steam inlet control valve, or governor, which is an isenthalpic expansion, it causes the turbine exhaust steam to dry out! This also reduces the amount of work that the turbine can extract from each pound of steam.

But, Dad! That's going to waste energy! You won't be able to get as much work from the steam! I can see how a reduction in the steam inlet pressure will dry out the exhaust steam, but it seems bad to me! Why not just avoid the whole problem by locating the turbine exhaust on the bottom of the turbine case? And then the water could just blow right into the surface condenser? See, Dad, how smart I'm getting! I'm already smarter than you!

Because then, the turbine would have to be elevated above the surface condenser, and this would be an additional structural cost!

 But isn't throttling the motive steam pressure to dry out the exhaust steam against some law? I think it's against the 2nd law of thermodynamics? Dad? Isn't it best to design equipment properly in the first place, by putting the turbine exhaust outlet on the bottom of the turbine case?

MAXIMIZING LMTD IN HEAT EXCHANGERS

Irene! You should always design a heat
exchanger for counter-current flow! That
means that the hot side inlet is next to the
hot side outlet of the other fluid! Also, it's a
convention to put the cold side on the bottom
of the exchanger!

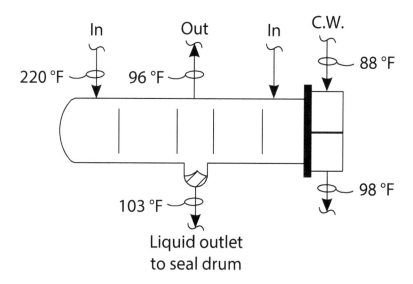

Vacuum tower
overhead vapors

Liquid outlet
to seal drum

Yes! I understand, Dad! Hot to hot, cold to
cold, to maximize the log mean temperature
difference! Okay! But how about your convention
of putting the cold side on the bottom? Is that
also one of your engineering rules?

Not really! I was working in Aruba at the
former Exxon Refinery. They had eight giant
vacuum tower overhead condensers! Both the
vapor inlets and outlets were on the top of
the shell, which is normal! All eight parallel
shells were that way!

So, Dad, I guess they put the cooling
water, tube side inlet nozzle on the
bottom of the channel head? Because
they're Exxon, they wanted to follow the
convention! Ha-ha!

Well, yes! The cold cooling water inlet was
on the bottom of the exchanger and the
warm water outlet was on top! That is the
convention!

But, Dad, then on your sketch, the warm
water outlet would have been next to the
vapor outlet? Didn't you write in one of
your dumb books that no one reads, that
in a vacuum condenser, the main thing is
to minimize the condenser vapor outlet
temperature?

Troubleshooting Process Operations is not
a dumb book! It's been in print since 1980.
But, yes, Irene, Exxon had made a mistake!
I had the cooling water flows reversed, so
that the cooler water entered on top of the
exchanger's channel head! This reduced the
condenser's vapor outlet temperature by
around 4°F!

Only 4°F colder vapor? You had them spend
thousands of dollars on the cooling water
piping changes for only 4°F! Boy! I bet they
were real mad at you! Did they kick you out of
the refinery? Expel you from Aruba? Have you
been arrested for fraud?

Well, Irene! The 4°F reduced the vacuum tower
top pressure by five mm of Hg. This resulted in
a reduction in the tower's flash zone pressure
of four mm of Hg! The resulting extra gas oil
recovery from the vacuum resid product which
was blended into #6 oil fuel oil sales was worth
$18,000 a day, which my client—

Wow! $18,000! Did they split that extra money with you 50-50? That would only be fair! It was your idea, Dad! If I had $9,000 every day, I could buy a new car every week! That would be great fun! Oh boy! I can't wait to get my chemical engineering degree! Wait 'til Gloria sees me driving around in a new car every week! My first car, I think, will be a silver Porsche 911—

CHAPTER FOUR

ROUTINE REFINERY OPERATING PROBLEMS

DR. PETRY & PAT

Me? I'm Pat. I grew up across the river. Been here at this plant 25 years. I'm just trying to keep her running!

Good morning. I'm Dr. Robert E. Petry, Senior Refining VP. So glad to meet you. Our corporate objective is to be the innovative leader in the refining industry.

PUMP BEARINGS LUBRICATION

 Pat, you are over-filling the bearing housing! Your oiler glass levels are full! Corrective action is needed at once!

Pump bearing lubrication

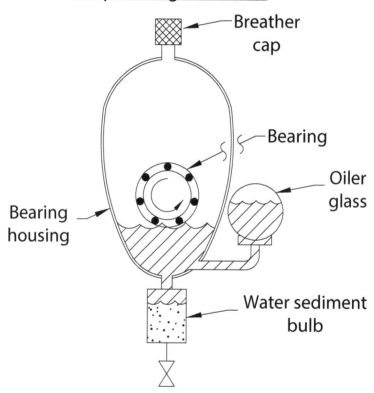

What'd you say, Captain?

Look to your Oiler Glass level, man! They're too full! The bearing housing oil level is much too high!

Actually they're not. They're mostly empty. I mean, the bearing housing is more empty than full, Bob!

The optimum lube oil level, as I learned at LSU, is up to the bottom race of the bearing. The level in this oiler glass is 2" higher than it should be! Are you listening?

That level ain't right in the oiler glass. There are three problems! Wanna hear them, Petry?

That's Dr. Petry! Okay, I'll listen for a minute. But be brief!

- First, looks like that breather cap's all dirty. The trapped air's gonna pressure-up the bearing housing, and push the oil up in that oiler glass!
- Second, there's water accumulation in the bearing housing! You see the water that I drain off the water sediment bulb?

- Third, the oiler glass is on the wrong side of the bearing housing! It needs to be on the left hand side for the direction of rotation!

Air? Water? How on earth could water and air enter the bearing housing? It's sealed up. That's what the lip seals or carbon seals are for!

Well, Doc, if the lip seals on the shaft, on either side of that bearing housing, are bad, the spinning of the shaft sucks in wet air! Round about here, in the River Parishes of Louisiana, the air's so humid, it condenses in the bearing housing. Screws up them bearings too! Here! I'll just drain the water off the bearing housing, clean the breather cap, and presto! You see, that oil level's disappeared in the oiler glass!

Very interesting! But I've got to go! I've got a full meeting schedule for the rest of the day!

From now on, I'd better stay in the control center! This sort of activity will damage my executive image!

Author's Note: Draining <u>all</u> water from the bearing housing does not fully restore the oil's lubrication properties!

PRESSURE MEASUREMENT PROBLEMS IN VAPOR LINES

Pat! We need to change your pressure gauge on the gasoline splitter overhead vapor line again! It's reading about 5 psi too low! I believe I had requested you to change this gauge in the morning! It's now well past lunch!

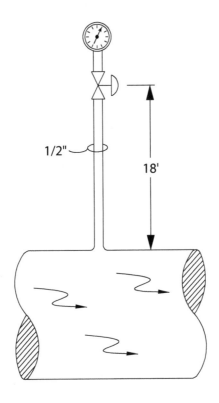

1/2"

18'

Look, I done changed out that dang gauge twice this week! There ain't nothin' wrong with it! I told you this before, Dr. Putrid!

My name's 'Petry'! Not 'Putrid'! The calibrated tower top pressure transmitter is reading 35 psig, just a few feet downstream of the gauge that's reading 30 psig! Your gauge is incorrect!

Okay, Your Majesty! I'm gonna explain this one more time. Then, you're on your own! The gauge is connected to a $\frac{1}{2}$" riser pipe, which fills itself up with light naphtha! Because the riser pipe is 18 feet above that vapor line, you gotta do this calculation:

- Head Pressure in Riser Pipe = 18 ft. × SG/2.31

Since the SG of the light naphtha is around 0.68, the head pressure liquid in that riser pipe is about 5 psi! You got it now, Your Lordship?

I hardly require you to explain elementary hydraulic calculations to me! But surely, since the $\frac{1}{2}$" riser pipe is straight and vertical, it will be self-draining? Hence, it would not fill with liquid, as your calculations pre-suppose!

Don't ask me why – but thin pipes, like $\frac{1}{2}$" or
$\frac{3}{4}$", don't drain! Don't matter if them liquids
are water, naphtha, or diesel! If the riser pipe
was two inches diameter, it would drain itself
down for sure! I'll show you! See, I'll take the
pressure gauge off, and blow-out the $\frac{1}{2}$" pipe
clear of liquid! Then, when I put the gauge back
on, it reads 35 psig! But then, it slips down
slowly, to 30 psig! Takes a couple of minutes!

Well Pat! You are quite correct! Your gauge is
fine! Just as you claimed! All very interesting
and informative!

It was a black day in my life, when
fate caused this miscreant Pat to
contaminate my world-view!

Pat! Perhaps you could reduce that 18 ft. riser
connection to 6", to clear up this pressure
discrepancy!

NEGATIVE PRESSURE DROPS

I suppose I am being overly critical, Pat! However, kindly observe that the shell side outlet pressure from the lead reactor effluent cooler, is 10 psi **larger** than the inlet pressure! Ha-ha! Of course, this is not physically possible! Kindly, Pat, redo your pressure survey more carefully!

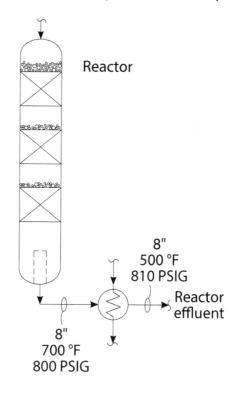

Reactor

8"
500 °F
810 PSIG

Reactor effluent

8"
700 °F
800 PSIG

I did it right! Just 'cause you don't like the numbers, Petry, that don't make them wrong!

But my good man! Both pressure connections are at the same elevation! Hence, the downstream pressure **must** be less than the upstream pressure, due to frictional losses in the exchanger! That's elementary technology! It's an irrefutable outcome of the Second Law of Thermodynamics!

I sure ain't your "good man," Petry! Didn't they learn you nothing in school about converting kinetic energy into pressure? What school you went to anyway?

I received my PhD from MIT! And yes, I am quite knowledgeable about the relationship between velocity and pressure! As the kinetic energy of a flowing fluid diminishes, its pressure will increase! Quite elementary physics!

Yeah! So, as the reactor effluent cools from 700°F down to 500°F, it partly condenses! Since the inlet and outlet nozzles both are 8", the shell side flow slows down a lot, and the exchanger outlet pressure goes up by itself! You understand that now, Your Lordship? Happens all the time out here!

Ah, yes! Just as I said! Energy must be conserved, according to Bernoulli's principle! It's gratifying to see a lay person, such as yourself, applying this important principle in everyday life!

Hmmm! Actually, I've seen this pressure rise before in plant data. But I always adjusted the recorded pressure profile data to make it appear more logical! Perhaps it is best to just report the data as observed in the field?

COKED-UP THERMOWELLS

Well, Mr. Petry – that coker feed drum temperature indication is reading too damn low! 'Bout 40-50 degrees off!

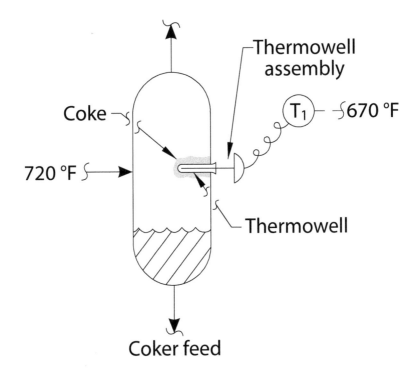

That's Dr. Petry – NOT Mr.! We have spoken about this before! Also, you're mistaken! I have recently directed the I&E Department manager to have that TI point recalibrated, to within 1°F. It's reading correctly!

Sorry, Dr. Pete! It ain't the thermocouple mili-amp output that's screwed up! It's that the thermowell itself, inside that there drum, is covered with coke! The thermowell is like a 1" piece of shiny stainless tubin'! Gets covered up with solid coke like stuff! Kinda like insulation round the thermowell! Then some of that heat gets radiated from the outside part of the **thermowell assembly**! Cools off the thermocouple wire junction itself, inside the thermowell! Reduces the electric output from the thermocouple junction! Bingo! A lower temperature readin'!

Well, I suppose you've employed your X-ray vision to see the coke? And my name is not Pete – it's Dr. Robert Petry! What then, Patrick, is the basis for your ridiculous and unfounded statement, that the internal portion of the thermowell assembly is fouled with coke? Where is the test to prove your quite outlandish theory?

Well Bobby! I done threw my coat over the outside part of that thermowell assembly! And the temperature readin' increased by around 30°F! That happened 'cause there's less radiant heat loss if I cover up that outside part of the thermowell and electric stuff! You know – less ambient heat loss! Wouldn't have happened that way if the inside part of the thermowell wasn't coked up!

Interesting observation! A rather clever field troubleshooting technique! How did you happen to learn to apply this method? Did they teach you this concept at LSU? Ha-ha! Ha-ha!

Hmmm…? Perhaps I can expand this concept! I'll submit it, with a few differential equations, as a manuscript, to the Journal of Advanced Process Control. But I had better not list Pat as a co-author, as he does not even have a PhD!

CENTRIFUGAL COMPRESSOR SURGING

 Pardon me, Patrick! Do you not hear the alky refrigeration compressor surging? Open up the discharge, to suction, **spill-back** immediately, to stop the surging!

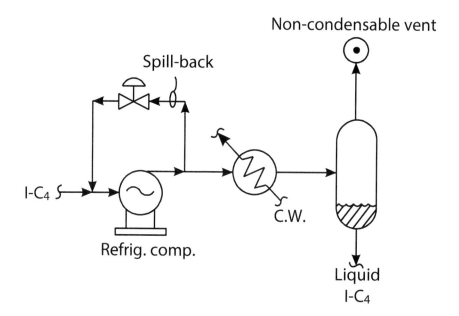

No can do, Doc! I'll make it worse!

Listen to me, you arrogant fool! The reason centrifugal compressors surge is that they are operating too far back on their flow vs. head performance curve!! Opening the spill-back will increase flow through the compressor and permit the machine to run-out on its curve, so as to stop the surge! Surging is dangerous and destructive! Even an uneducated and obtuse operator like you must appreciate this basic idea!

Yeah, yeah, Professor Pete! But you're forgettin' something! You're forgettin' the effect of density. Low-density vapors at that compressor suction reduce the pressure put up by that compressor. That low discharge pressure makes it harder for the compressor discharge flow to push into the downstream reactor. If that compressor can't climb up that high hill at its discharge, it kinda slips back down the hill! That's what surge means to me!

It's Dr., not Professor! But why would the density of the vapors decrease, when the spill-back is opened? It's just I-C_4? Certainly I agree that a reduced vapor density at the compressor suction would promote surge, but I do not quite follow your logic – if there is any!

That damn spill-back connection's on the wrong side of that there discharge cooler! It's spillin' hot $I\text{-}C_4$ vapors back to that compressor suction! Every extra 50°F at that compressor suction, reduces vapor density by around 10%, I figure! My pal, Norm, had that same problem back in Texas City in 1974 on an alky unit refrigeration compressor! He done told me about it when we went crawfishin' last month!

Very well, Pat! Let's not argue over this detail! Just increase alky feed to generate more cool vapors! That will definitely stop the surging!

Wonderful! I'll now propose a brilliant project to the plant manager! We will relocate the spill-back take-off from upstream, to downstream, of the compressor discharge after-cooler off of the top of the refrigerant vessel! I am sure that everyone will be duly impressed!

COMMISIONING STEAM TURBINE

 What exactly are you waiting for, Patrick? Let's get started! We need to commission the new steam turbine driven pump today!

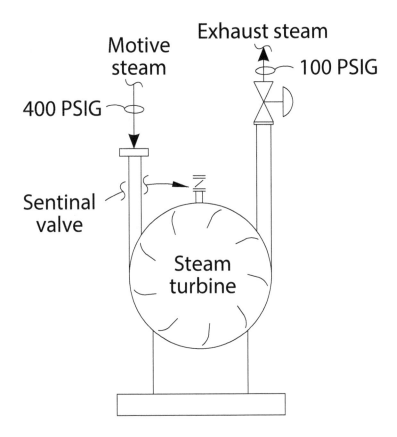

Okay, Your Excellency! But I need Johnny to help me open that 16", 100 psig exhaust steam valve first! Do you know where Johnny has got to?

Do not refer to me as "Your Excellency." John is off somewhere! Doubtless executing some misdeed! Please line up the 6" 400 psig motive steam, while we await his return!

Can't do that, Your Honor! I need to open that 16" exhaust valve first. If I open that motive steam 400 psig supply valve, with the 100 psig exhaust valve shut, the turbine case will pressure itself up to 400 psig. It ain't rated but for around 130#! Man! If I do what you say, I'll blow myself up! You can see, I'm too young to die! Ha-ha!

Do not call me "Your Honor," you dolt! You cannot overpressure the turbine! There is a relief valve located right atop the turbine case! Look! There it is! See it?

Oh! I beg your pardon, Your Lordship! But that ain't no relief valve! That's a **"SENTINEL"** valve! It'll whistle if the turbine case pressure gets much above 120 psig. It's just a reminder to always open the exhaust steam valve FIRST, and the supply steam valve last!

You know very well that my name is Doctor Robert Petry! Unfortunately, I do not have the hereditary title of "Lord." Regardless, I see that John is approaching. So, as we discussed, open the exhaust steam valve first, and the supply steam valve afterwards! Safety must always come first, Patrick!

We hear and we obey, Sir Robert!

I wonder if we cut Petry up, if he would make good bait for my crawfish traps? I'll ask Johnny if we can use some Petry pieces when we go crabbin' next week. He must be good for something!

REFRIGERATION SYSTEMS

Patrick! The refrigerant receiver vessel is beyond its service life! Let's retire it! It's pretty much corroded! It's really not required. We'll just control the refrigerant flow off the condenser itself! I'll write up the change order after lunch! It's just a wide spot in the line!

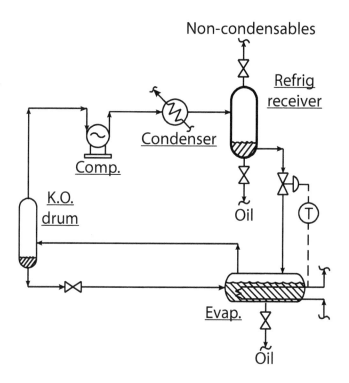

Man! I don't know! We done always had it! I kinda think we need it! I vote that we replace it!

We do not vote on process decisions in this plant! We employ technical judgment, based on logic and our technical training! That vessel is redundant! That is my final, executive decision!

Well then, Mr. Executive, how is we gonna keep that there refrig condenser from blowing through, and losing its condensate seal? That vapor will just blow on through that ole evaporator, and load up that refrigerant compressor!

My dear man! We will just keep the condenser TRC outlet valve throttled back a bit! See, there is no problem! Ha-ha! It's quite simple!

Okay, Your Lordship! But how's you gonna keep from losing a bunch of condenser capacity, due to condensate back-up, if'n I don't have that refrig. receiver? It needs to work together with that TRC control valve to keep that vapor from blowing on through the condenser, and also keep that refrigeration condensate from backing up in the condenser shell. The refrig. receiver and TRC work together like a team – kinda like you and me! Ha-ha, ha-ha!

Ah, yes! I must agree with your analysis, Patrick! They rather function as a steam trap! To prevent blow through, and condensate back-up, simultaneously! I am quite gratified that you agree with my analysis! The refrigerant receiver shall be duly replaced in the 2016 turnaround! Thank you for your input!

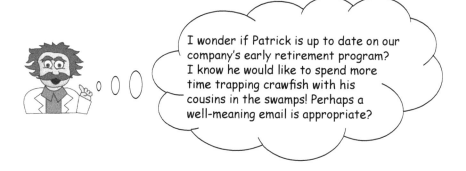

I wonder if Patrick is up to date on our company's early retirement program? I know he would like to spend more time trapping crawfish with his cousins in the swamps! Perhaps a well-meaning email is appropriate?

Author's Note – This incident occurred at the El Dorado Arkansas Viscous Polypropylene plant in 1968. I'm looking at my photo at the site, as I write this.

SEAL PAN DRAIN HOLE

Hey, Dr. Petry! I done rejected the splitter tower! I just told them scab contractors not to close-up the tower and tray deck manways!

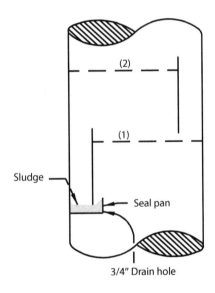

3/4" Drain hole

What? You can't do that Pat! The turnaround is already behind schedule!

What exactly is your problem? We need to get that tower ready for a Sunday start-up!

Look here, Petry! I told you yesterday that seal pan on tray #1 is full of crap! That's why we shut down! 'Cause the tower was flooding! Flooding 'cause of downcomer choking and backing-up 'cause of the dirt in that there seal pan! Even you can follow that, I reckon?

 Keep a civil tongue in your head! Have you then forgotten that I directed the maintenance supervisor, Bill Brundrett, to thoroughly clean the corrosion deposits out of the seal pan? Has this not been done satisfactorily?

Oh, yeah! The damn seal pan is nice and clean now! But Petry, I also asked you to tell your pal Brundrett to drill me a $\frac{3}{4}$" hole in the bottom of the vertical lip of the seal pan, to keep that water insoluble iron sulfide crap from building back up in that seal pan again. Where's my hole? I ain't gonna approve that tower for closure without that flushing hole in my seal pan!

I see your point, Patrick. Without a means to prevent sludge accumulation in the seal pan, we will flood the bottom downcomer again! And the flooding will back-up the tower! I will direct Mr. Brundrett immediately of our requirement for the $\frac{3}{4}$" flushing hole you have specified! Good work!

And you can also tell your pal Brundrett to tighten up all of them **DOWNCOMER BRACING BRACKETS** that keep the bottom edge of the downcomers at the 18" spacing from the wall of the splitter! All of them nuts and bolts on them brackets have done worked their way loose! Tell him to use double nuts this time! And stainless hardware! Not that carbon steel crap he buys on the cheap.

MEASURING STEAM FLOW WITHOUT A STEAM METER

 Patrick! Add 3,000 lbs/hr of steam to the diesel stripper! The flash is too low!

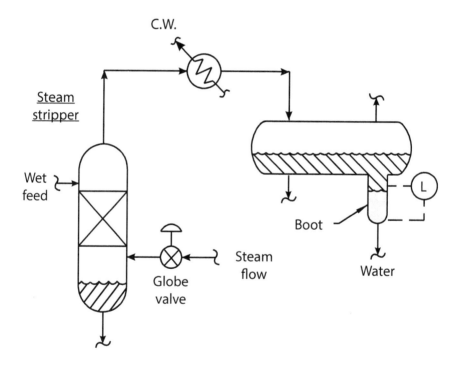

 Petry! There's no flow meter on the steam. How do you expect me to know how much is 3,000?

Hmmm... I have a brilliant idea, Pat! Shut off the boot pump! Then, shut off the stripping steam, and watch how long it takes for the level in the boot to rise by one foot!

I know all your ideas are brilliant – at least in your own mind. But, I don't get this idea at all, Petry!

I'll continue if I may! Next, open the globe valve controlling the stripping steam part way, and carefully note how far you've opened it! Then, shut off the boot pump again and see how long it takes for the water level in the boot to rise by the same one foot! And then...

Okay, I got it now, Bob! The difference in the boot's fill rate is the extra water from the stripping steam.

Patrick! That's Dr. Petry, not "Bob." All we need to know then, is the I.D. of the boot! You can perform my rather clever experiment at various globe valve positions, and thus create a permanent chart of globe valve position vs. steam flow rate!

Yeah! I guess that could work! Wanna help me do it?

Well! Every dog has his day! I guess even that idiot Petry has an occasional good idea!

Can't! I have an important luncheon meeting to attend with Mr. Duland and the new V.P.!

AIR LEAK ON SUCTION OF COOLING WATER PUMP

Patrick. The pressure on our cooling tower "B" pump seems 20 psi below normal! The water supply pressure flowing to the alky unit has fallen! Switch pumps! We can run on the "A" pump until the maintenance department can overhaul the "B" pump. My diagnosis is that the "B" pump will need a new impeller. Also, a new impeller wear ring!

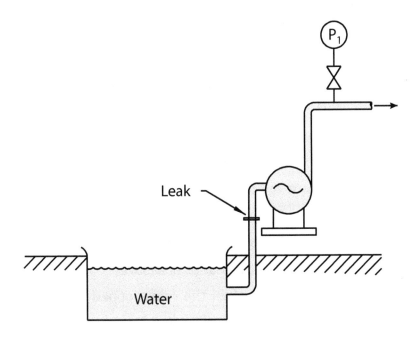

Yeah, well, Professor Bob, my diagnosis is that we need to paint the suction line with tar!

Tar? Are you joking? Come my good man! This is not the Middle Ages! Tar indeed! Further, I would prefer if you addressed me as Dr. Robert Petry, not Professor Bob!

Okay, Your Lordship. The problem is that air is being sucked into that pump's 14" suction line. Most of that line is running below atmospheric pressure. Like one or two psi below zero-gauge pressure! The sucked in air is making the pump slip and lose flow and discharge pressure! Kind of like cavitation, but not really!

But how do you know that? Patrick, how can you tell that air is being drawn into the suction line? Have you had the line X-rayed without my approval?

Easy, your Royal Highness! Yesterday it rained real hard! The "B" pump discharge pressure picked up by 5 psi. Then, when the rain stopped, the discharge slipped itself back down! I kinda duplicated the rain with a firewater monitor, and the same thing happened!

Rather clever, Patrick! But how about the tar? How will that help?

It will seal off the air leaks more permanent like, then just with the water! I used the same technique on my pool pump at home! Worked out real nice!

   ○ ○ ○   Even primitive man has a certain degree of native cunning needed for survival in the natural world! Instinct in action!

 Good analysis, Patrick!

CHAPTER FIVE

REFINERY SAFETY

DANGEROUS DAN & SAFETY SALLY

Don't worry! I always operate this way! Nothing has ever happened yet! I'm basically lucky! Life's a lottery!

Anything that can go wrong, will eventually go wrong! I like to analyze operating procedures! Maybe I worry too much?

AUTO-IGNITION OF HYDROCARBONS

 Gasoline has a low auto-ignition temperature! It's really dangerous to get a hot sample of light naphtha! Heavy products are safer to sample! Even if they're really hot!

Auto-ignition of hydrocarbons

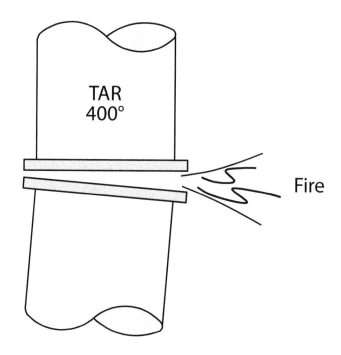

Safety first! But diesel has a lower auto-ignition temperature than gasoline! It's more dangerous to sample diesel at a high temperature than naphtha!

What, Sally? Everyone knows that gasoline catches fire more easily than heavier hydrocarbons – like diesel!

Sorry, Big Dan! You're getting confused between flash point and auto-ignition! Flash means the temperature at which a hydrocarbon ignites when exposed to a flame. The flash point of gasoline is room temperature and the flash point of diesel is above 150°F!

 What's auto-ignition mean, then? Isn't that kinda the same thing as the flash point?

No, Dan! Auto-ignition is the temperature at which a substance will catch fire by itself! The higher the carbon-to-hydrogen ratio, the lower the auto-ignition temperature! Professor Carl can explain. He's super smart!

Dan, the auto-ignition temperature of:

- gasoline is about 480°F (the lower the octane, the lower the auto-ignition temperature).
- asphalt or vacuum tower bottoms auto-ignition temperature, is about 300°F-350°F.
- natural gas might be 1,000°F or so.

Sorry to disagree, Professor. But I've sampled vacuum resid in a steel can at 500°F, and it didn't catch fire. I've done it for years, Sally, and no fires! Ha-ha!

Dan! Just you wait. One day you'll spill that resid sample on a porous surface – like your clothes! The vacuum resid will come into contact with a lot of air. Then you'll find out the real meaning of auto-ignition... DEATH! If something can go wrong, eventually it will go wrong!

FAILURE OF MECHANICAL PUMP SEALS

Dan! I'm worried about the mechanical seal integrity of our butane pump!

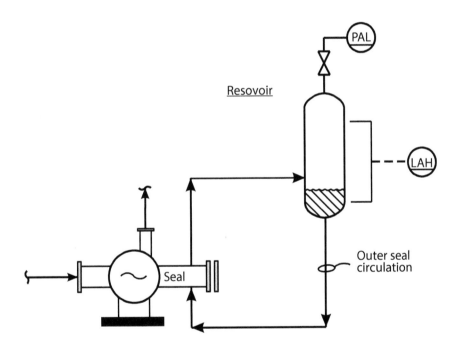

Sally! You worry too much! That pump has a double mechanical seal! We're double safe! The outer seal is a back-up for the inner seal!

No, Big Dan! If the inner seal leaks, then the high-pressure alarm on the outer seal reservoir should sound! The outer seal is not so much a back-up for the inner seal, but its function is to give the operator time to react to the failure of the inner seal!

Girl! Isn't that what I just said?

Daniel! When's the last time you checked to see if the pressure alarm on the reservoir worked?

Aw!! It will probably work ... I guess? You know, Sally, you can't worry about everything! I'll refill it on my next round, OK?

No! Both the high level and the high pressure alarms need to be checked routinely! That's the whole purpose of tandem seals! Alarms not checked routinely will never work in an emergency. Furthermore, I am not your "Girl"!

Sally's a great operator! But she should look at life through my "Rose Colored Glasses." We've never had a butane cloud detonation, even though we've had a couple of butane clouds due to our alky iso-pump seals blowing out. Sally – she forgets that we hardly ever have a source of ignition on our unit. Unless someone is doing hot work, or a truck drives through the unit, or we get a spark from some idiot dropping a tool, or we…!

SAMPLING TAR SAFELY

 Sally! We need to pull a vacuum tower bottom sample. Pronto!

Sample TAR safely

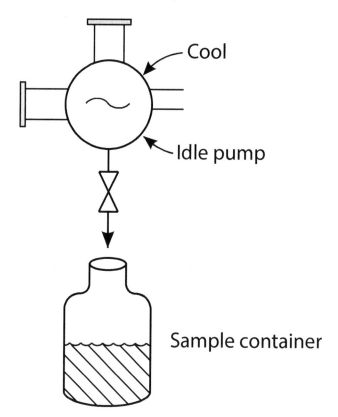

Cool

Idle pump

Sample container

Sorry, Dan! We don't have a sample cooler for VTB (asphalt).

Just get the sample in a metal can! Be careful to hold the can in the extension rod. Also, whatever you do, don't drop the can or spill any tar! That stuff is above its auto-ignition temperature! Anyway, sample coolers tend to plug-up in vacuum resid service!

Well, Daniel – there's a safer way to sample resid, without a sample cooler. Interested?

We always catch resid samples my way! And there haven't been any fires yet! So, what's your idea?

Pretty simple! First, switch pumps. Then, block-in the seal flushes to the idle pump. Wait until the pump case cools below about 300°F. Then, get your sample off the pump case drain! Below 300°F, the asphalt is below its auto-ignition temperature, so it's safe to sample!

That woman doesn't care about time. Sometimes, you've got to cut a few corners to get the job done!

Okay, fine! Do it your way, Sally! It's going to take five times as long – but who cares about time, if it's safer your way – Ha-ha! I'm going for a smoke!

DANGERS OF IRON SULFIDES

 Have you washed out that sulfur condenser yet, Sally?

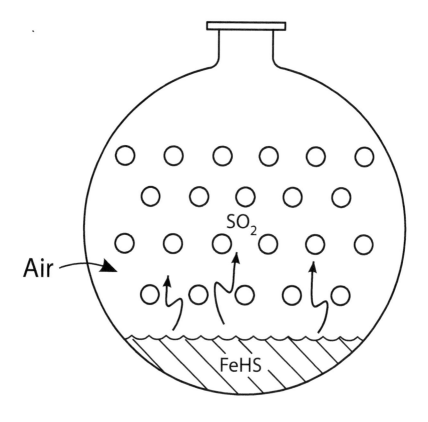

Yes, Dan! I washed it down with fire water!

Okay, good! Now issue an entry permit, after testing for H_2S, hydrocarbon vapors, and breathability!

Yes, Dan, I've done that already, but ...

Don't waste time! We're down for a four-day turnaround – that's it!

But Daniel! We haven't chemically cleaned the vessel to remove iron sulfides. They can auto-ignite at ambient temperature.

Well, woman, if you've washed out the vessel really good, you would have washed out those iron sulfides! Wouldn't you?

Not exactly, Dan! Iron sulfides ($Fe(HS)_2$) are not at all soluble in water!

So...? If the iron sulfides start to burn, they don't give off all that much heat. Let's not worry!

No, Dan! It's not the heat! It's the sulfur dioxide! SO_2! It's quite as deadly, in a confined space as H_2S. That's why we need to chemically clean vessels, to remove iron sulfide deposits, prior to issuing an entry permit!

Sally! Did you know that sulfur oxides are good for a man's virility? Ha-ha! Oh, okay! Order up a truck of Zyme solution and we'll chemically remove those naughty iron sulfides. Or even inhibited 10% hydrochloric acid will work! Maybe even citric acid?

Sally wastes more time on this safety stupidity than anyone else in this plant. She worries about every little thing. My motto is, "What? Me worry?"

H$_2$S FATALITIES

 Be careful, Sally! That rotten egg smell is H$_2$S - hydrogen sulfide! It's a deadly gas - especially to females! Ha-ha! Ha-ha!

H$_2$S smells bad = safe

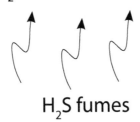

H$_2$S fumes

H$_2$S no smells = death

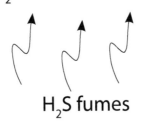

H$_2$S fumes

You're certainly correct, Big Dan! H$_2$S is a killer! But it's more dangerous when you don't smell it, than when you do smell it! Also, unlike certain unnamed people that I work with, H$_2$S does not discriminate between people, regardless of race, sex, or national origin.

What? The more concentrated the H_2S, the worse it smells. That's common sense!

Not true, Dan! At low concentrations, like 20 or 50 ppm, H_2S has that rotten egg smell. But it's not dangerous! At 1,000 ppm, the H_2S deadens the sense of smell. A few breaths will knock you off your feet. You won't smell anything! And then you're – well, permanently dead!

Are you sure about this, Sal? Is this some kind of theory you got from Norm? You know he makes a lot of stuff up as he gets older! He's a bit senile!

Norm told me he has lost friends in Aruba and Texas City, 'cause of H_2S. In Aruba, it was rich amine that flashed. In Texas City, the acid gas backed out of the flare header, when they stupidly pulled a reboiler bundle, without blinding away from the flare line!

Ha-ha! Sally! The flare header is almost always under a small sub-atmospheric pressure! What Norm told you was hogwash!

Hmmm.... Now that I think about it, the flare header pressure can go a coupla inches of water positive. But I hate to admit to Sally that what I just said is not 100% correcto!

FIRED HEATER – POSITIVE PRESSURE

Sally, open the air register on the heater! The excess O_2 is only $\frac{1}{2}$%!

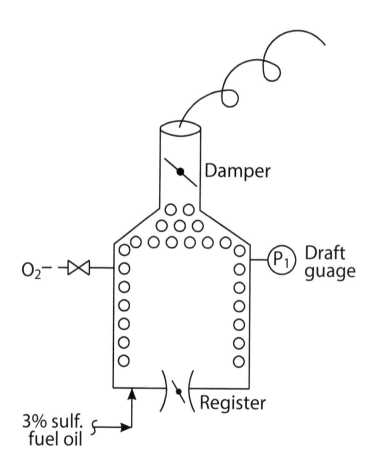

Perhaps that's not too safe, Dan?

What's not safe, is running a natural draft heater air deficient! Just add more air Sally – NOW!

But Mr. Dan, we may go positive at the draft gauge located below the convective section! That might be dangerous?

That's why I'm called "Dangerous Dan"! Ha-ha! A little bit of positive pressure for a while can't hurt anything, Sally!

But, Dan! If we go positive below the lower row of convective tubes, we will blow SO_2 out of the convective section of the box. Don't forget we're burning 3% sulfur fuel oil! Not natural gas!

Hmm... SO_2? Well, it will all blow away! Let's not worry too much! It's windy today!

Sorry to disagree! But Norm is working on the heater taking draft measurements! And SO_2 fumes are deadly! Let's open that stack damper first, to keep a negative draft of like 0.1 inches of water below the bottom row of convective tubes!

SO$_2$? Well, that SO$_2$ will be good for Norm's virility! Ha-ha! He should not be out here in the plant anyway! Engineers belong in their office!

Wow! This may be a great opportunity to get rid of that annoying engineer! Why can't Norm stay in the office like the rest of management? It will just be an accident! Heh! Heh!

REFINERY EXPLOSIONS & FIRES

 Accidents always can happen, Sally! It's in the hands of God! Hope for the best is what I always say. Look on the bright side of life!

Refinery fires & explosion

150 mils/yr corrosion	10 mils/yr corrosion
Carbon steel	410 chrome steel

Both brown, rusty color
both magnetic

I really have to disagree, Big Dan! Every accident that has happened at this refinery since I hired on in 1997 could have been avoided! Plan for the worst is what I think!

That's ridiculous, Sally, and you know it!

Well, Daniel, how about the time that carbon steel spool piece blew out on the coker circulating reflux and the gas oil auto-ignited! The line was supposed to be 410 S.S.! The corrosion rate for C.S. at 600°F and TWO% sulfur was 180 mils per year!

 Sally, that was 'cause some idiot insulated over the C.S. pipe, and it couldn't be easily inspected! It was just a short spool piece!

And remember when we burned down the visbreaker fractionator!

 Sally! Clearly an Act of God! There was no fire proofing around the fractionator's skirt! Don't you believe in God?

Sure! But I remember God gave man free will, and the power to reason! You remember Jimmy?

Yeah! Good ole boy! Sorry he had to go that'a way!

Yeah, Dan! Blew up with H-103! The crude heater! The heater got itself caught up in a positive feed-back loop! You should have switched the fuel to manual, and backed-off on the fuel CV – not sent poor Jimmy out there to open the air register and the stack damper and then to...!

 Hold on, woman! I've heard enough outta you for today! Ain't it 'bout time for you to check your pumps? This here plant ain't in Bhopal, India, you know! This ain't no Union Carbide Pesticide Plant.

 That's what comes of sending Sally off to that Norm Lieberman Safety Seminar in New Orleans last year! Nothing but trouble!

FLOODING FIRE BOX WITH FUEL

 Sally! I asked you before lunch to close those air registers on H-101. What's holding you up, girl? We got too much O_2!

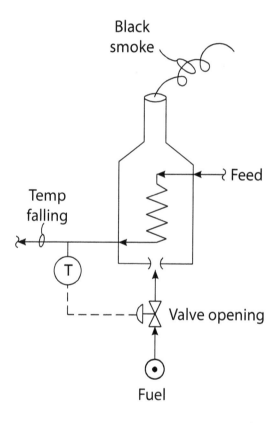

The firebox looks a bit hazy, Daniel! It might be dangerous to cut back the combustion air flow! Also the flame tips look a little bit long and smoky!

Look here! The stack O_2 is $4\frac{1}{2}$%. Our target, according to Mr. Overbourne, is 2%-3%. I just had the O_2 analyzer recalibrated Friday! It's still okay! We got lots of room to cut air to that heater! Let's get to it, woman!

You know something, Dan? You're a real fool! We're running that heater on auto-TRC! Suppose our heater inlet temperature drops by 20°F! Then what? Do you know what will happen, Mr. Macho Man?

Sure! The fuel gas control valve will open to restore the outlet temperature to its set point! That's what the TRC is supposed to do!

And suppose we become a bit air deficient? You know, Dan, if air-fuel mixing is bad because of tramp air leaks, then the heat release in the fire box will drop, as more fuel is added, even with 2% or 3% excess O_2 in the stack! What's going to happen then? What do you think?

Well, I guess since we're on auto-TRC – the fuel gas CV to the burners is going to open more! But without more air, this won't help! I guess it's going to reduce the heater outlet temperature even more! I guess the heater's gonna get caught up in a positive feedback loop!

And the heater outlet temperature will just keep on dropping! And the firebox will get more and more fuel rich. And then we'll have to...!

We'll have to rush outside to open the air registers again! And maybe, also the stack damper! We're going to need lots more air! That's for damn sure! More air, quick!

No! No! No! You'll wind up killing yourself! Dan, the first thing to do is put the fuel on manual control! Then, manually reduce the fuel gas rate! The heater outlet temperature will go up at first! But when it starts to go down – only then – is it safe to increase the air flow by opening either the stack damper or the air register! Then, switch back to auto!

But our instructions, Sally, say to target between 2%-3% O_2 in the stack! You think that you're smarter than the engineer who wrote those instructions? Smarter even than Mr. Overbourne himself?

ROUTING RELIEF VALVES TO THE FLARE

Big Dan! This plant is really dangerous! All
our alky pressure relief valves are vented
to an atmospheric vent. Some direct to the
atmosphere! Some to a blowdown system that
then vents to the atmosphere. I'm worried!

Routing relief valves

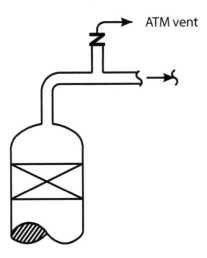

ATM vent

You're worrying about nothing again, Sally!
Those pressure relief valves are all located
on the vapor line, or on the top of the tower!
We only got vapor at the top of the deprop,
debutantizer and the naphtha splitter! If
that relief valve pops open, it's going to vent
just vapor to the air! And those vapors will
just drift away! Not a problem!

How long you been out here, Dan? How many times you see that naphtha splitter FLOOD? Well, Dan?

 Hmm...? I hadn't thought about flooding. I guess then we might just puke light hydrocarbon LIQUIDS out of the blowdown atmospheric vent, and from the relief valves direct to the air! That's kinda bad! But I guess it's still okay, 'cause we don't have any sources of ignition round here!

And how about all those scab contractors driving their pick-up trucks all around our unit? Don't forget what happened at the B.P. Refinery in Texas City a while back! Killed 15 people and injured about 300 others. Cost to B.P. $5,000,000,000!

 You know something, Sally, you worry too much! Try to look on the bright side of life. Kind of like me! Ha-ha! You ain't gonna live forever anyhow!

I still say that all those relief valves should be tied into the refinery flare knock-out drum. Dan, better safe than sorry!

ISOLATING EQUIPMENT WITH GATE VALVES

 Will you please block-in that heat exchanger, Sally! The maintenance guys will soon be here to blind-off the tube side!

Isolating equipment with gate values

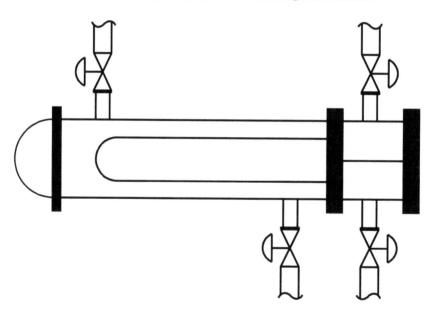

Okay, Dan! I'll block-in the shell side too! We'll need to blind that shell side as well as the tube bundle! For safety!

That's not necessary! We only plan to hydro-blast the tubes! The tubes are not leaking! So it's only necessary to blind-off the tube side, and not waste our time blinding off the shell side, as it won't be exposed in the atmosphere! Understand that, Sally girl?

And suppose, while hydro-blasting, we cause a tube to start leaking? Then what, Daniel? You need to try thinking ahead for a change! Think defensively!

Then what? Well – we've closed the isolation gate valves on the shell side! So even if we have a tube leak, the 480°F gas oil on the tube side, will not be able to leak to the atmosphere! Sally! Did you make sure to pull up nice and tight on the shell side isolation gate valves? Those are really big valves for a little woman to close! It's really a big man's job! Ha-ha!

Don't you worry! I used a two-foot cheater bar! The valves are pulled up real tight! But suppose there's some trash caught up in the gate valve's seat? Then what? The tube will leak and the valve will leak! And then hot gas oil above its auto-ignition temperature will leak into those tubes and then pour out the channel head! It's happened before! At American Oil in the 1960s. A big fire resulted. Also, sometimes gate valves leak across the top side of the gate! Then, pulling up harder can make the valve leak worse! We need to blind-off the shell side also, Big Dan! I'm worried!

Why do we have to assume both a tube leak and a leaking gate valve? I think that's called "Double Indemnity" or something? Why can't you ever take a more hopeful attitude, Sally? You're always so negative! Why can't you learn to be a regular guy like Carl or Pat?

Because I'm not a "guy." And also Daniel, we should always "Hope for the best, but plan for the worst!" By the way Dan, Gloria told me you dropped $400 at the casino on your vacation last month! I guess then, you were also hoping for the best. Ha! Ha!

EXPLOSIVE LIMIT OF HYDROGEN

Sally, I was just wondering…if you get a vessel that's full of air and gas – like hydrocarbon vapors – which is better? To have lots of gas or a little gas? Do you know? Which is more likely to blow up and kill me?

<u>Explosive limit of hydrogen</u>

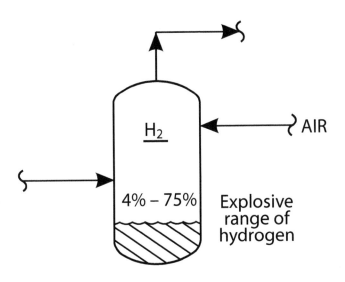

Dan, if you have a really high concentration of hydrocarbon in air, even if you have a source of ignition, it won't explode! The stuff inside the vessel can be too rich – or too lean – to explode! Like more than 10% propane mixed with air won't explode! 'Cause it's just too rich!

So, I guess we're pretty safe, as long as we're outside the "Explosive Range" for light hydrocarbons! I guess it's the same for all hydrocarbons – methane, ethane, propane, and such?

It varies, Daniel! But what you really need to be careful about is hydrogen! Like the off-gas from our naphtha reformer! Hydrogen rich gas has a really wide range of concentrations over which it can explode! But only if there's a source of ignition!

We ain't got any ignition sources here. So I can forget the whole business! But I suppose you'll worry about it anyway, Sally!

Daniel! Of course we have a source of ignition in all our process vessels! How about all the iron sulfides? They'll auto-ignite at ambient temperatures when they dry out! Think, Daniel! That's the key to safety! And how about static electricity – that's a source of ignition too!

By the way Daniel, did you know that when hydrogen burns – like a flange leak – it's mostly invisible to the human eye! That happened at a Chevron Plant near San Francisco back a while!

Yeah, Sally! But not to your eyes! You know everything!

Actually, Daniel, I looked up the explosive range of different gases in air:

- Hydrogen – 4% to 75%
- Methane – 6% to 14%
- Propane – 2% to 10%
- Butane – 2% to 8%

You can see, Dan, why I'm super careful when working with hydrogen! It's such a big explosive range!

CLIMATE CHANGE

 Sally! This stuff we hear about climate change is just plain stupid! Man! Last winter was super cold around my house! It's all a bunch of liberal, commie propaganda!

Climate change

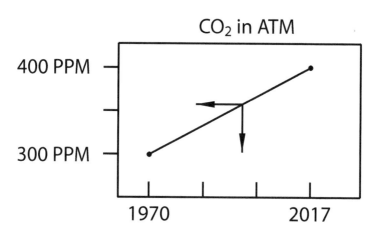

It's true, Big Dan, that we've had a cold winter here in the U.S.! But worldwide, 2016 was the warmest year on record! The world's heating up! The glaciers are melting! The Antarctic ice sheet is coming apart at its edges! August 2016, was the hottest month in the recorded history of our earth!

Sal, You worry about everything! Don't you even know that the ice sheet in the Arctic is getting bigger the past few years! Ha-ha! See, that proves global warming is just a liberal, communist lie!

That's also true, Dan! But the Arctic ice sheet is getting progressively thinner! I suppose this means the Arctic ice cover is getting more vulnerable to breaking apart? But the main thing, according to Norm, is that the CO_2 in the atmosphere has been increasing at a dead steady rate of between 0.5% and 0.6% every year, since the 1970s! Pretty much, that means CO_2 in the atmosphere will double in 100 years! That's certain to accelerate global warming! The history of the earth proves it for sure, Dan!

Woman! You're a fool! In a hundred years, we'll both be dead, gone, and forgotten! Why worry? We got us enough problems in the plant today! Are you all done checking the bearings on P-304? If we let them vibrations build up, we'll blow the pump seal Then we'll have a nice cloud of white, butane vapor, drifting over to No. 3 CAT CO boiler! Now that's somethin' real to worry about!

Right, Dan! I'll check P-304 right now!

I guess most folks think like Daniel. Can't blame him! He's got to worry about his kids playing video games and taking drugs, not global warming. Gloria's no prize either!

DANGER OF CARBON STEEL PIPING SPOOL PIECES

 Miss Sally! Did you know that carbon steel corrodes a lot faster than chrome steel? Like about ten to twenty times faster if the pipe's really hot and full of high sulfur oil?

Danger of carbon steel

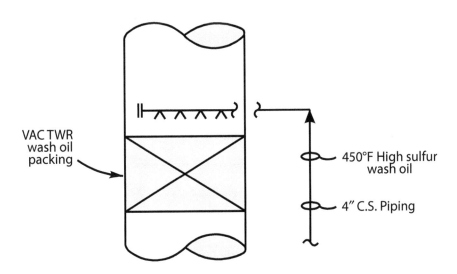

VAC TWR
wash oil
packing

450°F High sulfur
wash oil

4" C.S. Piping

Yep! I know that, Big Dan! Like 450°F and one or two % sulfur. At a refinery in Louisiana, they thinned a vacuum gas oil line running at over 600°F, to the point of failure, in less than a year! Blew out and caught fire! Nobody killed though.

What they replace it with Sally? 410 chrome? Bet they used 316 stainless?

Carbon steel! Ha-ha! That's a true story, Dan! It corroded down again to discard thickness in ten months! They didn't even realize it until Norm and Allen Hebert asked them to check out the pipe's metallurgy in their records! Carbon steel! Real fools!

Man! Those guys must have been blind! All you gotta do is look at a pipe to see if it's carbon steel or chrome! Chrome is gonna look shiny!

Not really! Both carbon steel and chrome steel both look the same! Kinda dull, brownish! You're thinking 'bout 300 series steel! It's shiny 'cause of its Nickel content! You're thinking about 304 or 316 or 317 stainless pipe.

Yeah! That's right! But you can use a magnet! Carbon steel, it's magnetic – but chrome steel ain't! Carbon steel has a magnetic personality – like me, Sally!

No, Dan! You're wrong again! Chrome steel is also magnetic! It's the 300 series stainless steels that ain't going to attract a magnet. You've got to look at the code stamped onto the flanges to figure out if the pipe's carbon steel or chrome!

Okay! Okay, Sally! I guess it's best to get these guys from Inspection to figure out if some line is nine chrome or just carbon steel! Man! I hate to think what would happen if that 700°F asphalt line blew itself out again! That was a really bad fire we had last summer!

 Working around Sally makes me real nervous! Can't relax! It's getting so bad, I'm startin' to look forward to going home to my wife and kids after each shift!

SCREWED CONNECTIONS

 Will you please remove that broken gauge and replace it with a new gauge! I've asked you to do this twice today! Exactly when are you going to get this done, Sally?

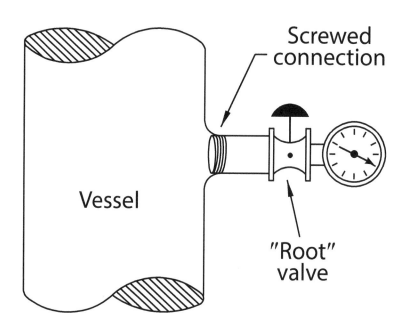

Daniel! When I try to screw out the bad gauge, both the root valve and the screwed fitting at the vessel wall also turn!

Really, Sally! Just get a back-up wrench! Hold the valve in place with the big wrench, and turn out the pressure gauge with the other wrench! You need me to help you?

No! I don't need your help! What I need you to do is think! An activity you frequently fail to engage in!

Look, woman! I'm trying not to argue with you! We need a new gauge on that reactor! Let's try to get along this week! It's bad enough working graveyards without arguing too! What's your problem?

The $\frac{3}{4}$" nipple is screwed directly into the vessel wall. When you thread a piece of pipe, you cut away half of its thickness to make the threads! So, if the nipple breaks, it's gonna break where it's threaded!

Sally! We got threaded connections all over this plant! All our fittings are threaded! How could you ever change a pressure gauge, unless you screwed it onto a threaded connection?

Okay, Big Dan! I'm just upset 'cause the threaded connection is between the reactor vessel wall and the "**Root Valve**." That's the valve next to the vessel! Our company piping spec's call for all connections between a vessel and a valve nearest the vessel – which I call the "root valve," must be flanged or "**back-welded**"! Suppose that threaded connection breaks? Then what? We'll have 1,200 psi, 700°F diesel blowing out! We won't be able to stop it!

Never mind! Go make the coffee! I'll change the gauge myself! Sorry to have bothered you with my little problems, Your Ladyship!

I'd better make a list of all the threaded connections inside the root valves, and have them all back-welded during the next turnaround! Otherwise, if one of those threaded connections break off, I'll never hear the end of this from Sally!

Author's Note: From an incident at the Exxon Refinery in Baton Rouge, #9 Pipe Still crude tower, 1985.

DANGERS OF STEAM DEAERATORS

Do you know, Daniel, that a contractor was killed at a chemical plant a few years back? Boiled to death! Norm was telling us about this, when he was teaching his Safety Seminar in New Orleans last year!

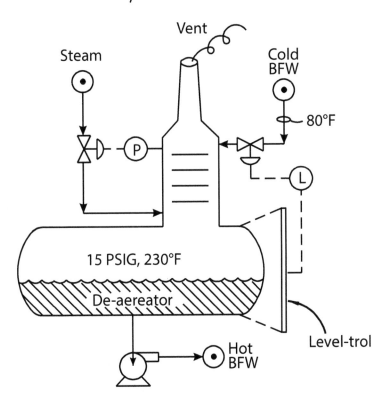

Wow! New Orleans! Did you go down to the French Quarter? They got some hot strip clubs down on Bourbon Street! But how could anyone get hurt from a Deaerator? Don't seem too probable!

Happened in Beaumont or Lake Charles. The problem was the steam inlet pressure control valve went 100% open. Boiling water blew out the atmospheric vent! The hot water landed on top of the poor guy! Third degree burns all over! Took him about a week before he died. Just awful!

I guess there must've been a failure of the cold water LRC valve! That's what overfilled that deaerator! Should've had a high level alarm; or trip; or something on that BFW level!

No, Dan! Here's what happened...it was all caused by the steam inlet pressure control valve being too small to heat up all that make-up 80°F BFW to 230°F!

Hold on, Sally Girl! You just said that steam was on pressure control! But now you're saying it's actually temperature control? Make up your mind, woman!

Lord Preserve Us, Daniel! You sure are dumb!
Water boils at 15 psig, or 230°F! It don't
matter none whether you hold 15 psig or
230°F, it's all just boiling water! The problem
is that once the steam – TRC or PRC – control
valve goes 100% open, the temperature at
the water's surface will get below 230°F,
and the deaerator pressure will start to fall.
And then:

1. The bulk of the water below the
   surface will boil up.
2. The boiling water will expand, and raise
   the real level in the deaerator.
3. The density of the boiling water will go
   down.
4. The indicated external level will also
   go down, because the density of the
   boiling water between the level taps, is
   gonna go down!

I reckon that the cold water feed valve is
gonna open up further and then...!

And then, Dan, since the steam valve is 100% open, the deaerator pressure and temperature are gonna drop like a rock inside your empty head! The real level inside the deaerator will rise above the steam inlet! Then, boiling water will blow out the vent stack. Caught up in a positive feed-back loop! That's what killed that contractor fellow!

CONNECTING STEAM HOSE TO HYDROCARBON SYSTEM

Excuse me, Daniel! You've just connected a steam hose directly to the isobutane 6" supply line to the reactor! That, Dan, is not safe!

Connecting steam hose

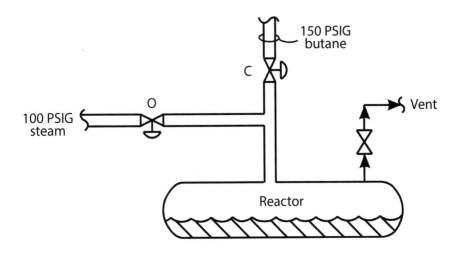

Don't bother me! I'm working! Don't you have some important meeting to go to? I'm trying to steam-out the reactor to get it ready for maintenance! We've got to give them an entry permit tomorrow! Remember, we got an Alky turnaround going on?

But Daniel! Let's take an extra minute to think safety! The steam is 100 psig! The 6" iso-butane line can pressure-up to 150 psig! What's to stop the butane from backing through the hose into the 100-pound steam header?

Look, woman! Use your eyes! The 6" iso-butane line is blocked in at the manifold! It's de-pressured! Go cook some chicken soup and stop harassing me, Sally! I need another length of steam hose! You can get it out the back of the switch room!

And just suppose, those idiots on the next shift open the 6" butane valve at the manifold! Then, what's to stop the liquid butane from backing into the steam header? Don't forget Frankie, your dumb fishing pal is working nights on "B" shift!

Yeah, Frankie! Maybe I better put a **Check Valve** in the connection between the 6" process line and the steam hose? Boy! Just imagine what would happen if we got a bunch of iso-butane liquid in the steam header, and some poor bastard used steam on the Cat to put out a fire! Wow!

Yes, Big Dan! A check valve is needed when we connect a process system to a utility system! Once at Tenneco Oil, they somehow cross-connected plant air to natural gas, and killed 16 contractors inside an aromatic splitter tower during a turnaround. Happened in 1980 or...!

Okay, don't start making speeches! Run down to the storehouse and get me a 1" check valve! And don't gossip with Gloria in the store room! Get me a cold drink while you're down there too. Pepsi or Coke, it's all the same to me! And a pack of donuts!

Yeah! I guess that woman is right! You got to be real careful when you connect up an air, gas, water, nitrogen, or a steam hose to a process vessel! Imagine what would happen if you cross-connect nitrogen to instrument air, like Norm once did on our unit! Our panel instruments blew out pure nitrogen, instead of air into the central control room! The idiot Norm almost killed us all!

PROCESS VESSEL COLLAPSE UNDER VACUUM DURING START-UP

 Did you steam-out the coke drum yet, Sally? We've got to air-free it! Steam it out really good for twenty minutes at least, to blow all the air out of the top vent!

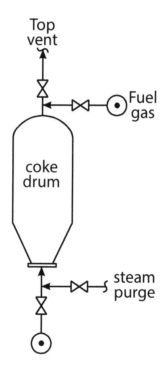

Yes, Dan! I did that! The coke drum is air free! I'm going out now to shut off the steam and line up the fuel gas to the coke drum! It'll take about half an hour. I'll be back for lunch then! Okay?

No Sally! I'm hungry now! Block-in the steam vent and the 100 psi steam purge, and let's go! Let's not mess around with the fuel gas...we'll get to that later...after lunch!

But Dan! I can't just close the vent and block-in the steam! That's dangerous! The coke drum could collapse! It happens all the time!

Collapse? Have you lost your mind, woman? That coke drum has one- inch-thick steel walls! Did you think that it's made out of plastic? Ha-ha!

It happened in Citgo! I think in Lake Charles! Those guys steamed-out a 20 ft. coke drum, blocked it in, and went off to get them a cup of coffee! The drum cooled off, the steam collapsed, and a vacuum developed inside the drum!

Yeah, I do kind of remember that! But what caused the drum to collapse anyway? There wasn't any pressure pushing it in, was there Sally?

Dan! This happened on the planet earth! There's atmospheric pressure! If the drum cooled off to 100°F, then the vapor pressure of water inside the drum was only one psia. The outside pressure was $14\frac{1}{2}$ psia. That's a big, big delta P. What I don't understand, Daniel, is why your head doesn't collapse, as there's also a vacuum there?

Okay, woman! Go ahead and break the vacuum with the fuel gas! I'll wait! I think it's all a big waste of time! But I don't like to argue with women! Just remember that I'm hungry! A man's got to eat to keep up his strength!

ACID GAS K.O. DRUM

 Excuse me, Sally! We have a high-level alarm on the acid gas K.O. drum! Could you see what's wrong? I don't want to carry a slug of water into the sulfur plant!

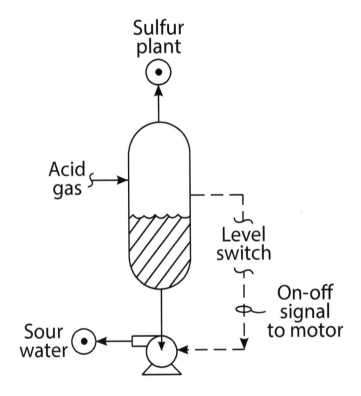

I guess I know the problem! The pump suction filter is fouled! Happened on the night shift last week! I'll get on my Scott Air Pack and get over there in about 15 minutes! I'll clear it, Dan!

Sally! Could you just not waste time! It's only sour water! Can't you just clean the pump suction filter? That water won't smell that bad! With the type of guys you've been dating, you should be used to some bad smells! Ha-ha!

My social life is none of your business! Anyway, I know exactly what that water will smell like! It won't have any smell! It's saturated with H_2S! My first breath will deaden my olfactory nerves. My second breath will knock me out. Then, I'll just lie there and die from hydrogen sulfide inhalation! It's happened before!

You missed out on a good career as a dramatic actress! Where do you get your crazy ideas from? Maybe from your *Ladies Home Journal* magazine? Just get on with it, woman!

Daniel! Did you know a while back, a female operator working at the Exxon plant in Baytown in the 1980s was killed clearing the suction of an acid gas knock-out drum – just like ours! Her shift-mates went out looking for her. Found her dead next to the pump with the suction drain valve open!

Okay, okay! Go get your Air Pack on! If something happens to you, I'll spend a month of Sundays filling out safety reports. I don't know why I ever even try to argue with you, Sally!

If something were to happen to that Ol gal, I sure would miss her! Best darn operator out here.

SAFETY NOTE

When opening a drain in H_2S service, it's a good idea to screw on a "Spring Loaded Dead-Man's Valve," which will close as soon as the operator releases his grip. Exxon purchased these after the incident related above. You can buy one from the "Mine Safety Catalogue."

PART II

Some subjects are a bit too complex for our little friends to explain. So, I have gone into further detail in Chapters 6 – 10. I hope if you have questions on these subjects, you will email me at: norm@lieberman-eng.com.

CHAPTER SIX

PITFALLS IN COMPUTER MODELING

A refinery model, based on the operating characteristics of process
equipment including:

- Distillation Trays
- Packed Towers
- Heat Exchangers
- Air Coolers
- Fired Heaters
- Steam Turbines & Surface Condensers
- Centrifugal Pumps – NPSH
- Compressors
- Piping Systems

can be created by computer simulation calculations. Superficially, the use
of computerized calculation methods are more accurate in representing
actual equipment functions than the older methods that have now been
largely displaced by modern computer calculation process engineering
techniques.

BASIS FOR PROCESS ENGINEERING CALCULATIONS

In the early 1960s, methods of executing process engineering calculations were gradually replaced by current methods. These rely on numerous iterative calculations that could never be done using manual calculation tools. In the early 1960s process engineering calculations were based largely on:

- Rules of thumb.
- Extrapolating from operating prototypes.
- Data obtained from performances testing.
- Charts and graphs extracted from plant testing.
- Shortcut methods derived from observations of plant data.
- Application of heat and material balance.
- Vapor-liquid equilibrium calculation.
- Application of the ideal gas law.

These 1960s methods were not suitable for any completely new process. But in the refinery industry, we have no new processes. Refinery process units are rather the same in 2017 as they were in 1960.

Thus we pose the question: which method of process engineering calculation is best for producing a cost-effective design for revamping an existing refinery process unit, or even the construction of a new refinery process facility:

- Modern-day computer simulation technology
  or
- 1960 manual methods

UNDERLYING ASSUMPTIONS IN DISTILLATION TECHNOLOGY

In order to design a distillation tower, the relative volatility of the components must be known. The definition of relative volatility is based on:

- Vapor pressure of the lighter component, divided by the vapor pressure of the heavier component, at a particular temperature.

The relative volatility is derived from the equation of state specified in the distillation computer simulation model. The user must select the appropriate equation of state from several dozen choices. Based on the equation of state selected, the following parameters are then calculated by the distillation computer model:

- The number of theoretical stages
- The reflux rate
- The reboiler duty
- The tower diameter
- The condenser duty

However, since the underlying assumption of which equation of state to select is typically made arbitrarily, these apparently rigorously calculated values are no more than an educated guess. However since many designers simply use the default value for the equation of state, perhaps the phrase "educated guess" should be revised to just "a guess."

TRAY FRACTIONATION EFFICIENCY

To calculate the number of actual distillation trays needed, based on the computer model's number of theoretical stages for a specified split, also requires the specification of tray fractionation efficiency. There are a number of published methods to calculate tray efficiency. Unfortunately, these methods do not take into account the real underlying factors that control tray fractionation efficiency. These are:

- Tray deck levelness
- Weir levelness (see Figure I)

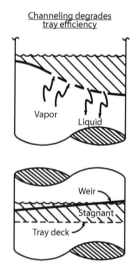

Figure I

The above statement regarding tray levelness is less true for bubble cap trays. However, bubble trays are now rarely used. Perforated trays are the norm. Perforated trays are:

- Sieve holes
- Valve caps
- Grid
- Tab tray without weirs

Based on mechanical design features, tray out-of-levelness can be anticipated, as can weir out-of-levelness. But without consideration of the relevant mechanical design features, tray fractionation efficiency is largely unknown and unknowable.

HEAT EXCHANGER TRAIN PERFORMANCE

Predicting the performance of a forty shell and tube heat exchanger train can be a complex task. Certainly, in a refinery crude preheat exchanger train, the service that I'm most familiar with, the controlling factor for predicting crude preheat is the crude side fouling factor, which ranges from 0.001 to 0.040 (reciprocal of BTU/HR/FT^2/°F). The fouling factor is a function of:

- Tube surface temperature.
- Velocity (i.e., fluid shear at the tube wall).
- Di-olefins and air that polymerize upon heating to form gums.
- Sediments, corrosion products, and other particulates.
- Tube roughness.

There is a definite, but variable rate of fouling, depending on mass flow velocity across a heat exchanger surface. High velocity, especially in viscous services, is limited by excessive pressure drop and erosion. What then is the correlation between fouling factor and velocity? It largely depends on the surface roughness of the tubes, the sediments in the flow, and the potential gums (i.e., di-olefins and oxygen) content of the process fluid.

But how can one quantify the effect of velocity and the other parameters on overall heat transfer coefficient? Without detailed experience in each particular service for a particular tube bundle configuration, the answer is unknown and unknowable.

HEAT EXCHANGER PRESSURE DROP

Anyone who has provided technical support on a crude unit, especially when the crude flow is on the tube side of the exchanger train, realizes that exchanger delta P is, after six months to a year after start-up, largely a matter of fouling. Observed pressure drop may be:

- Five times the clean delta P for crude preheat trains with low (2 ft/sec) tube side velocities after one year.
- Slightly above clean delta P for crude preheat exchangers with high (10 ft/sec) tube velocities, after three years.

How can one predict actual pressure drop, with crude on the tube side, with this variable relationship between clean and actual pressure drop, due to fouling?

The problem is even more complex with the shell side flow. On the shell side, complex internal clearances often determine observed pressure drop. Perhaps the seal strip configuration in older bundles does not represent the current seal strip configuration. Also tube sagging distorts the tube pitch. Gaps between the tubes due to the channel head pass configuration may cause shell side by-pass gaps that should, but often are not, sealed with dummy tubes.

How then, short of measuring heat exchanger pressure drops in the field and then extrapolating the DP in proportion to flow, can the designer predict the crude preheat exchanger pressure profile? The answer is unknown and unknowable.

PACKED TOWERS

Thin, flexible, aluminum rings are a recipe for disaster. The problem is crushing during installation. Premature flooding and excessive HETP (Height Equivalent to a Theoretical Plate) is to be expected. Beds of structured packing (not grid) are preferable.

A typical HETP for a bed of structured packing, with 1" crimp size, might be 20". Or it might be 60"+? It all depends on how well the liquid (and to a lesser extent the vapor) is distributed. The designer would use the packing vendor's correlations. The difficulty arises due to the liquid distributors:

- Not installed absolutely level.
- The distribution holes plugging.
- The distributor not bolted-up correctly.
- Fouling between layers of the packing.

What then is a reasonable factor to increase structured packing HETP due to these problems? More often than not, I've observed structured packing with a double HETP, than that calculated from vendor correlations. How can we correctly represent a bed of structured packing in distillation service in a computer simulation model? Short of field testing, the answer is unknown and unknowable.

And even if we could observe an HETP at 10,000 BSD of reflux, how do we extrapolate this to a rate of 20,000 BSD? Engineering judgment. That is, one has to guess.

And, if it's all going to come down to guessing based on experience, judgment, and extrapolation, what purpose do detailed computer simulations models really serve?

AIR COOLERS

For a forced draft air cooler (Figure II), the discharge air flow from the F.D. fan splits into two parts:

- Cooling Air
- Recirculation Air

The sum of these two parts remains constant, even when the tube fins foul. But even when the fins are new and clean, the percentage of air recirculation may be 10% or 20% or more. And when the fins foul, the percentage of air recirculation may be more than the air flow through the bundle.

Consult an air cooler data sheet. Has the designer taken this factor into account regarding heat transfer efficiency? Obviously not. How then can one correlate the loss of air cooler efficiency, vs. fouling, if the original design does not even take this obvious factor into account? And how will this factor due to fouling vary with time? The answer is dependent on experience gained from field testing. If the air cooler is used as the overhead condenser on a distillation column, then this variable air flow limit must somehow be factored into the computer model representing the distillation tower simulation. How can this be done, without reference to engineering judgment – or an educated guess?

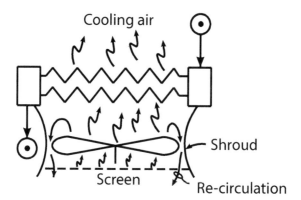

Figure II

FIRED EQUIPMENT

What is a reasonable target for computer control for excess oxygen, manipulating combustion air, to a natural or forced draft heater?

a.   0.5% minimum
b.   2% – 3%
c.   5% maximum

The answer is that there is no answer. At one plant in New Orleans, Louisiana, optimum combustion conditions, based on maximum energy economy was realized at 0.5% excess O_2 measured above the radiant section tubes. On an older unit in Port Arthur, Texas, with poorly maintained burners, and a leaking fire box, the optimum was 5% - 6%. It all depends on the air-fuel mixing efficiency of the burner. This mixing efficiency is a variable depending on:

- Tramp air leaks
- Draft balance
- Wind direction
- Composition of the fuel
- Use of premix burners
- Use of staged, low Nox burners
- Burner tip pressure drop
- Burner tip fouling
- Firing rate
- Percent of design firing rate

The optimum excess O_2 varies with time. How frequently will the optimum excess O_2 change? It depends on how variable the process. The answer as to what is a reasonable excess O_2 at any moment cannot then be correlated with any O_2 analyzer. Placing the O_2 analyzer above the convective tube banks magnifies the effect of convective section tramp air leaks.

How then can we use modern firing control to optimize the combustion air rate? Not with an O_2 analyzer, or with computer input. Does this also imply, that O_2 analyzers, placed above the convective tubes in a natural draft heater, will serve no purpose? In my experience, such is the case.

PIPING SYSTEMS

When we calculate hydraulic losses in piping systems, using a computer simulation of the piping network, it is necessary to input:

- Line lengths
- Fitting (elbows and valves)
- Flow
- Fluid properties
- Piping friction factor

The piping friction factor is a function of the roughness factor. Which the user of the piping program must guess. Having guessed at the roughness factor, the resulting calculations, while done with precision, are also engineering approximations – or just guesses.

What then is the correct way to calculate piping pressure losses, if computer calculations are not to be believed? The engineer must field check the observed DP in a similar service and ratio the flow based on:

- Delta P varies with velocity, squared.
- Delta P varies inversely with pipe diameter, raised to the 5th power.
- Delta P varies linearly with line length.

But suppose the engineer does not have reference to the relevant field DP measurements? Then how to calculate pressure drops with precision? The answer is unknown and unknowable.

CENTRIFUGAL PUMPS – NPSH

For proper pump design, "The available NPSH (Net Positive Suction Head) should be equal or exceed the NPSH required." This common statement is not true. If one wishes to improve the mechanical seal reliability, the available NPSH must be double or ten times greater than the required NPSH. But how much greater? It depends on how tolerant you are to mechanical seal failures. Figure III shows a rough approximation of excess available NPSH vs. required NPSH.[1] Exactly what is meant by increased reliability is somewhat undefinable. However, a computer simulation, calculating a vessel height to satisfy centrifugal pump NPSH requirements, will not take this into account.

Providing NPSH somewhat in excess of a pump's required NPSH will not stop vaporization inside the pump case. It just reduces it to a few percent. What then is a reasonable excess of available over required NPSH? It depends upon the individual circumstance as to how dependable the pump must be.

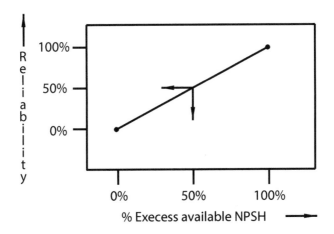

Figure III

ROTATING EQUIPMENT

The efficiency of steam turbines varies with the relative pressure drop across the governor speed control valve (which is bad) and the pressure drop across the port nozzles (which is good).[2] The efficiency of a centrifugal compressor will vary from 60% to 85%, depending on rotor fouling. The efficiency of a reciprocating compressor will vary from 0% to 90% depending on the condition of the suction and discharge head-end and crank-end valve plates and springs.

Assuming some computer generated, or standard efficiency, is not an engineering approximation. It's an unsubstantiated guess, without a field observation of the calculated gas temperature rise, between the suction and the discharge temperatures, to verify the assumed efficiencies.

SUMMARY

What's wrong with computer modeling of process units? Nothing! Computer modeling itself is fine, if used as an aid to interpret field data and observations. Computer models, if based on a prototype, are a useful <u>EXTRAPOLATION</u> resource. But computer modeling, if used as a stand-alone calculation tool, without the support of plant data, does not and cannot represent the reality of process equipment performance.

The problem with process design using simulation models is that they are a distraction from the real job, which is the same in 2017 as it was in 1965 – or even 1935. That is, the need to base process equipment calculations on performance testing, direct plant field observations, laboratory data, and the experiences of plant operations personnel.

REFERENCES

1.   Bloch, Kenneth. "Rethinking Bhopal," 2016. El Sevier Publications.
2.   Lieberman, N.P. *A Working Guide to Process Equipment*, 4th Edition. McGraw Hill.

Value of a Chemical Engineering Degree to the Process Engineer in a Refinery

Most of what we need to do our jobs as Refinery Tech Service process engineers, we learned in high school. The time we spent in university seems, in retrospect, to be a waste of time and money. We were simply not instructed in the methods needed to solve process problems, which is pretty much what we have been hired to do.

The key to refinery profitability is not market conditions, or management, or operator expertise. It is effective process engineering technical service. As we were not taught this in university, could we learn to become effective process engineers by learning from older guys we work with? In 1965, when I started work for American Oil, this was the method I used. But, it won't work now. Too much knowledge has been lost. You're on your own now.

So let me suggest you learn by running performance tests. For example, some of the tests I have run, from which I learned a lot, were:

- Optimizing pumparound circulation rates.
- Varying steam pressure to vacuum ejector.
- Changing tower operating pressures.
- Varying stripping steam rate to gas oil stripper.
- Increasing pitch angle on blades to air cooler.
- Slowing down turbine driven pump.
- Opening adjustable head-end clearance pocket on reciprocating compressor.
- Varying liquid levels on circulating thermosyphon reboilers.
- Effect on fractionation efficiency with incremental reflux.

Having run such tests, see if you can interpret the results, using your technical education. That is, can you calculate the effects you observed? Most of what I know about process engineering was derived in this way.

Computer simulations and seminars are pretty much like university. They are too remote from reality. Reading books has helped me, but until I try to practice what I have read, it's not all that helpful. It's kind of like most of life; we learn only by doing.

CHAPTER SEVEN

LATENT HEAT TRANSFER

Exactly what is the benefit of going to high school … I cannot quite understand. I imagine it's an opportunity to meet girls? For example, in high school Chemistry Lab, I was instructed to add a **Boiling Stone** to my flask before setting the flask of water above my Bunsen burner. But why? What would happen if one omitted these magical stones? After all, my mom boiled water in our apartment's kitchen all the time, without the aid of boiling stones.

However, six years later, I learned to respect the function of boiling stones. I was working as a tech service engineer on a crude distillation unit for American Oil in Whiting, Indiana. I had drawn a sample, in a quart glass bottle, of light naphtha from the crude tower reflux drum, which operated at 15 psig, or about one bar. It was really cold in Whiting that day. And my sample bottle was brand NEW!

I planned to submit the naphtha sample to the refinery lab for distillation analysis, and had left it open on my desk to allow the light-ends to weather-off, to check the percentage loss of the naphtha before sending it on to the lab. After an hour, the bottle had warmed to room temperature, but none of the naphtha had evaporated. I recall taking a glass rod from my desk drawer to stir up the sour naphtha – which then exploded out of the bottle into my mouth, nose, and eyes!

So, I had to go back to the crude unit reflux drum for a new sample. But this time, I dropped a small pebble into the glass bottle to provide a site for **Nucleate Boiling**. Which is, of course, the function of the boiling stones we used in high school. Then, as my naphtha sample warmed, the naphtha gradually vaporized without incident.

Years passed. I was working for the Coastal Refinery in Aruba. In my hotel room, I had a new microwave oven and new teacups. I decided to have a cup of tea. I noticed, however, that the water failed to boil in the microwave, even after ten minutes. Quite odd, I thought, as I set the very hot, still full cup on the sink's counter. But as soon as I dropped the tea bag into the cup, the water exploded out of the cup with great violence.

What happened? Lack of Nucleate Boiling Sites! The new tea cup had no scratches on its interior surfaces. But my tea bag had suddenly provided many such sites.

PROPANE-BUTANE SPLITTER REBOILER

A few years later, I was having dinner in a cheap Mexican restaurant in Texas City with Brent, the tech service manager of the local Valero Refinery.

"Norm," Brent began, "I just had a new tube bundle installed in our propane-butane splitter. The old bundle had 10% or more tubes plugged, so we had it re-tubed. It turned out really badly. Now, we can barely get half of the old reboiler duty. The 100 psig steam flow is down by half. I can't understand it. What's happened? Our plant manager is real mad, and blames me!"

"Brent, a consultant's time and advice is his stock and trade."

"Yeah, Norm. Okay. I'll pay for dinner."

"You know what your problem is, Brent? Lack of Nucleate Boiling Sites. Like in high school we used boiling stones in chemistry lab. You remember? When you replaced your tube bundle, you threw away all the dents and scratches on the outside surface of the old tubes. You discarded all of the existing Nucleate Boiling Sites. Especially, when you are vaporizing a mostly pure component, such as water, or in your case, normal butane, essentially all of the heat duty is in the form of latent, rather than sensible heat transfer. See where the bubbles are forming in your coke. The CO_2 bubbles are forming at the scratches in the bottom of the glass. Those are the Nucleate Boiling Sites. If you drop some sugar crystals into your coke, the rate of CO_2 bubble formation will greatly accelerate. If I order a beer, the rate of..."

"Norm. You're getting off the subject! What should I do to restore the capacity of my C_3-C_4 splitter reboiler?"

"Brent, I'll give you three options:

1. Buy a new bundle with a **Sintered Metal Coating**. It's a porous metal, electro-plated type material, deposited on the tube bundle. Kind of expensive though.
2. Lightly sand-blast the new bundle to roughen its exterior surface.
3. Put back the old bundle."

Brent selected option #3, and it worked out just fine. Brent also selected not to pay for our dinner, as he claimed, "Consultants customarily pay for dinners with their clients." Which is true, when the consultant is actually paid by the client, and not being tricked into supplying technical advice for free!

HEAT FLUX LIMITATIONS

Donald Kern in his excellent book, which I use frequently, *Process Heat Transfer*, McGraw-Hill (1950) on pg. 459, states that the maximum allowable flux rate for natural circulation reboilers and vaporizers, that are vaporizing organics (such as hydrocarbons) is 12,000 BTU/HR/FT^2. For forced circulation, it's 20,000 BTU/HR/FT^2. The "FT^2" is the exchanger surface area.

The problem with Mr. Kern's values for maximum flux rate is that I ordinarily see them exceeded by a large amount in refinery distillation tower services. Yet, in some circumstances they appear to be quite a valid limitation. Actually, this subject of heat transfer flux limitations began for me when I was 13 years old.

I had decided to go into business for myself. My business was making quarter-sized lead slugs. In the 1950s, vending machines only checked the size of the coin. I had found steel washers with an I.D. that exactly matched that of a 25-cent piece. I found some tire balancing lead weights on the street. Using my mother's new, shiny frying pan, I melted the lead, and poured it into my 25¢ molds. I used the newly minted coins to buy milk (25¢ a quart) and cigarettes (but I did not smoke). My coins did not work, however, in the local telephone booth.

But one day, while heating my mom's pan to melt the lead, I accidentally dropped some water on the very hot pan surface. Contrary to my expectations, the water did not instantly flash away to steam. To my surprise, the globule of water danced merrily on the surface of the shiny new pan in quite a lively manner.

It didn't seem to be evaporating very fast – if at all. What was happening? Stranger still! Only when I entirely shut off the gas to the burner and waited a few minutes did the glob of water collapse, spread out across the surface of the pan, and instantly flash away to steam.

When I was 13, I had no friends. Even my mother did not particularly like me. "Norman," Mom would say, "Why can't you be more like your sister, Arline? She's so friendly. Everyone likes her!"

And I would say, "Mom, don't bother me. I'm thinking."

I was thinking about water dancing on a hot frying pan. You can see why I had so few friends as a teenager. I was thinking about a heat transfer, flux limited situation. Why was it that a droplet of water took five minutes to evaporate on a very hot surface, but only a second on a moderately hot surface? Also, I noted one day, that on one of Mom's older, scratched frying pans I could not reproduce the dancing water globule behavior. The water would instantly evaporate, regardless of the temperature of the frying pan.

And what was the outcome of all my experiments and efforts? The wicked vending machine company revised their cigarette, candy, and milk dispensing machines to check coins for electric conductivity and weight, so that my lead slugs were rejected. But 20 years later, there was another result of my teenage observations. I had designed a revamp of an existing isobutane-normal butane splitter for Amoco Oil at their Texas City Refinery. As part of the project, I had replaced the existing reboiler tube bundle with a new bundle. The old bundle had 1" tubes on a rotated square pitch. My new bundle had ¾" tubes on a triangular pitch and hence 20% more heat exchange surface area. I required the larger surface area because the new service of the tower was to fractionate between Iso-pentane and normal pentane. This raised the reboiler process side temperature about 50°F:

- $Q = U \cdot A \cdot Delta\ T$

Q (the duty) was the same. But the Delta T between the 50 psig steam and the boiling pentane was smaller. Hence the increased required A (Area).

This was a natural circulation reboiler, rather than a forced circulation reboiler, hence according to Mr. Kern's book, the maximum flux rate should have been 12,000 BTU/HR/FT². But the actual observed flux rate prior to the revamp was 21,000 BTU/HR/FT².

So I decided to ignore, to my sorrow, Mr. Kern's 12,000 flux limit. When we started up, with the new reboiler tube bundle, I could only get about 60% of the previous reboiler duty. And, sad to say, the calculated flux rate was 12,000 BTU/HR/FT².

I decided to switch to higher pressure, 100 psig steam, which condenses at a 45°F higher temperature than the 50 psig steam. And the reboiler duty went down further! In my mind, I slipped back to my aborted childhood criminal career, making lead slugs. I recalled how the droplet of water danced around, cushioned by a thin layer of steam, from physical contact with the hot, smooth frying pan. I recalled that the hotter the pan, the longer it took to evaporate the glob of water.

"Ah!" I thought, "A flux limited situation. Just as Mr. Kern predicted."

And then I remembered about how the water would evaporate much more readily on the old, scratched, cast-iron pan, than the new, shiny, stainless pan my sister, Arline, had given Mom as a gift.

We ordered a new identical bundle. We had it sent to Union Carbide for an external coating of sintered metal, **High Flux Tubing**, and the required design reboiler duty was readily achieved.

I have seen the flux limits that Mr. Kern states for maximum heat flux exceeded many times. But I have also seen the 12,000 flux rate limiting reboilers and vaporizers. What did Donald Kern do wrong? Nothing! He likely had data or experiments based on smooth, new alloy tubes. Most of my refinery work is with old, pitted, partly corroded carbon steel tubes.

In summary, there is a very definite relationship between a nucleate boiling limitation and a heat flux limitation, in the sense that both problems are rectified by using a roughened, heat transfer surface.

Having experienced these problems as a teenager helped me to understand the underlying nature of boiling heat transfer coefficients. It's true my social skills are somewhat retarded as a result. But my big sister Arline says she has noticed some definite signs of improvement since I turned 74.

CHAPTER EIGHT

HYDRAULICS

Have you ever wondered what a water tower is for? I used to think it was to store water. But if that was the water tower's only function, why have the water tank elevated 100 feet above grade? Why not just locate the tank on the ground?

The reason that the water tank sits 100 ft. above the ground is to maintain a constant pressure in a town's water supply. If I fill a glass tube 28 inches or 2.31 feet high with 60°F water, and place a pressure gauge at the base, I will read a pressure of one psig. Or, I can say, 2.31 feet of water exerts a head pressure of one psig. If the water tower's tank is 100 feet above my house, then:

- 100 feet ÷ 2.31 = 43.3 psig

That's the water pressure that I'll have in my home in New Orleans. I'll use this pressure to water my vegetable garden. The water flows through my garden hose at a pressure of 43.3 psig. When I say this, I'm ignoring frictional losses; I'm assuming head losses due to the roughness inside the hose and the city water piping system is zero. At the end of the hose I have a brass nozzle. The pressure of the water coming out of this nozzle is less than 43.3 psig. The pressure of the water at this point is zero psig (or 14.7 psia at sea level in New Orleans).

What has happened to the energy represented by the 43.3 psig of pressure? In reality, some of it is converted to friction. But I have decided to assume frictional losses are zero. Therefore, the 43.3 psi reduction in water pressure has been converted into:

- Velocity (feet per second)
  Or
- Kinetic energy ($\frac{1}{2}$ m • V^2)
  Or
- Momentum (mass times velocity)

I'll be using these three terms somewhat interchangeably in this text. According to Dan Bernoulli's equation:

- Delta P $\propto$ (Velocity)^2

Delta P means pressure drop. The ($\propto$) symbol means proportional. For example, if I double the water flow in a pipe, the pressure drop in the pipe would increase by a factor of four. (Incidentally, Dan's dad was the first person to credit Isaac Newton for creating Calculus, the source of unimaginable student suffering.)

Actually, I can calculate the nozzle exit velocity (again assuming no frictional losses) as follows:

- Delta P = 0.18 (Velocity)^2 (1)
- Delta P = 43.3 psi = 100 ft = 1200 inches
- 1200 = 0.18 (Velocity)^2
- (Velocity)^2 = 1200 ÷ 0.18 = 6670
- Velocity = (6670)$^{\frac{1}{2}}$ = 81 feet per second

The 0.18 factor is derived from the gravitational acceleration constant of 32.2 feet per second squared, here on the surface of the planet earth. If you are visiting my home planet from a remote galaxy, kindly correct the above calculation for the force of gravity on your own home planet.

Now, let's assume you point the water nozzle at the end of your hose, straight up. Can you guess how high the water will rise above my garden at a nozzle exit velocity of 81 feet per second, assuming no air resistance or wind or friction? The correct answer is … 100 feet, the level of the water tower.

I've converted velocity, or kinetic energy, or momentum, into another term of Daniel Bernoulli's equation. That term is Potential Energy. I have converted the 81 feet per second nozzle exit velocity into 100 feet of potential energy.

I've forgotten to mention that I have a tank elevated above the roof of my house, 100 feet in the air. When I check the pressure of the water in this tank at grade, it reads 43.3 psig.

FLUIDS OTHER THAN WATER

The calculations I have performed assume that the water s.g. (specific gravity) is 1.00. Other liquids have a different specific gravity, or s.g. For example:

- Hot water at its boiling point = 0.97
- Kerosene or jet fuel = 0.80
- Gasoline or naphtha = 0.70 to 0.75
- Diesel oil or No. 2 oil = 0.85
- Butane = 0.55
- Propane = 0.50
- Mercury = 13.6
- Liquid gold = 25
- Air (in New Orleans) = 0.001

Therefore, if propane was placed in a tank 100 feet high, it would create a liquid head pressure of:

- $(43.3) \bullet (0.50 \div 1.00) = 21.7$ psi

If I squirted the lower pressure and lower density liquid propane vertically into the air, it would rise to an elevation (ignoring the fact that the propane would in reality all vaporize) to 100 feet. The nozzle exit velocity would be the same as for the water, that is, 81 feet per second.

ROTATIONAL ENERGY

My dad bought my first bicycle for me 65 years ago, for my 10th birthday. Unfortunately, the bike was big and I was little. So, instead of riding the bike, fulfilling its intended function, I decided to perform an experiment. I placed my bike upside down on its seat and handle bars, and squirted water against the front wheel's spokes. The wheel spun around with a large rotational speed. At the time, I thought that it was the pressure of the water that was spinning my bicycle's wheel. But now that I'm older (but not actually wiser), I realize it was not the pressure striking the bicycle wheel's spokes that was providing the rotational energy, but the velocity, or kinetic energy, or the momentum of the water.

One day, rather than risk my life riding my new giant bike, I developed a project. My plan was to place a pulley on the bicycle wheel. The spinning pulley would transmit its rotational energy to a rubber drive belt, which would then spin a centrifugal water pump. The water that I had used to spin the bicycle wheel, having now lost all of its momentum (mass times velocity), could then be pumped into a water tower 100 feet high. And my very clever plan would have worked fine, except for two minor problems:

1. Friction.
2. My dad took my bike away until I learned to ride it.

PRESSURE DROP THROUGH AN ORIFICE

One morning, when I still lived with my parents in New York City, I was taking a bath. Suddenly, my mother banged on the bathroom door.

"Norman," she called, "What are you doing in the bathroom so long? Your father has to go to work."

What I was doing was conducting a fluid flow experiment. I had placed a plastic sheet with a single half-inch hole over the bathtub drain. Over the hole in the plastic sheet, I had placed a plastic cup with ten half-inch holes, to prevent the formation of a vortex from interfering with my experiment. Then, I had turned on the bathtub faucet and waited until the water level in the bathtub had stabilized. That is, the water flow from the faucet equaled the water draining from the tub. I then measured the flow from the faucet in a one-gallon milk container.

Using this observed volumetric flow and the half-inch orifice size, I calculated the nozzle (i.e., orifice) exit velocity. Next, I calculated, based on the steady state, or equilibrium level in the bathtub the, "Orifice Coefficient," (i.e., the 0.18 shown in equation (1), above). But it was not 0.18 but closer to 0.3. What had I done wrong?

Actually, nothing. The difference between 0.18 and 0.3 coefficients was due to friction or turbulence. Most of the 14 inches of liquid head in my bathtub (that is, 60%) had been converted to velocity:

- $(0.3) \cdot (60\%) = 0.18$

But the remaining 40% had been lost due to friction or turbulence. Meaning,

- $(40\%) \cdot (14 \text{ inches}) = 5.6$ inches of the water's potential energy had been converted to heat. I suppose the water temperature increased by a fraction of a degree, as it accelerated through the one-inch orifice in the plastic sheet. This is called, "Head Loss."

EFFECT OF FLUID DENSITY ON ORIFICE PRESSURE DROP

Note that I've expressed the total conversion to both acceleration and into friction in terms of the 14 inches of liquid head, or potential energy, in my bathtub. If I had bathed in kerosene or mercury, the conversion of potential energy (that is, the 14 inches of liquid times the 60%) into acceleration would have been the same. The frictional losses (14 inches of liquid times the 40%) would have been affected by the fluid's viscosity. However, in my work as a refinery engineer, I deal almost totally with low viscosity fluids, so that the conversion of liquid height to velocity is very seldom affected by liquid's density or viscosity.

Pressure loss is another matter. Pressure drop or loss is directly proportional to density. I can relate pressure loss to head loss as follows:

- Delta P = (Delta H) (Density)   (2)

  Or

- Delta P = $\dfrac{\text{(Delta H) (S.G.)}}{2.31}$

Where

- Delta P = Pressure drop, psi.
- Delta H = Head loss, feet.
- S.G. = Specific gravity (i.e., density relative to cold water).

To summarize, when I say I'm converting pressure or head to velocity, it's really not a loss. Ideally, I can get that pressure or head back when the fluid slows down. But when I say I'm converting pressure or head to friction, or heat, or temperature, then that head is pretty much gone forever. I'll discuss this point further in my chapter pertaining to "Thermodynamics."

HEAD LOSS IN PIPING

What does it mean when we read, "The head loss was four feet per mile of line?" Let me explain.

I own a ten-acre farm in Louisiana. Near my house is a fish pond. Five hundred feet away (0.1 miles) is the edge of the swamp. The pond is six feet above the elevation of the swamp. I've dug a ditch for the pond to overflow into the swamp. After it rains, I'll watch the water run through the ditch and think, "I'm losing six feet of head in my ditch that is 0.1 miles long. Or, I'm losing:

- (6 ft.) ÷ (0.1 miles) = 60 ft. of head per mile of ditch

If I replace the open ditch with 500 feet of four-inch PVC pipe, then I'll lose 60 feet of head per mile of line. However, is this line loss due to increased velocity or friction?

If the water is running through a section of pipe at a constant velocity without any change in the pipe's elevation or diameter, then the head loss is entirely due to friction. If the interior of the pipe is smooth, the friction loss is small. If the interior of the pipe is rough, then the frictional losses are larger. In order to calculate the pressure drop in a pipe then, we have to know the degree of roughness inside a pipe. A factor that is normally not known. But I can calculate the head loss with this equation:

- $\text{Delta HL} = \dfrac{(0.4) \bullet V^2}{(\text{I.D.})}$ \qquad (3)

Where:

- Delta HL = Head Loss due to friction, per 100 equivalent feet of piping, in feet.
- (I.D.) = Pipe inside diameter, inches.
- V = Velocity, feet per second.

A 100 feet of straight pipe is the same as 100 equivalent feet of piping. Every elbow, U-bend, and valve increases the number of equivalent feet by creating turbulence in the flowing fluid.

The 0.4 coefficient is an empirically derived value. That is, I have worked in refinery and hydrocarbon process plants for 53 years. In general, the 0.4 coefficient is a reasonable value for process piping that I have observed for water, gasoline, diesel oil, kerosene, and other low viscosity fluids. For any fluid with a viscosity above ten centistokes, or above 50 SSU (Saybolt Seconds Universal), a more detailed calculation is required.

The accuracy of this calculation is plus or minus 50%. Meaning, if I calculate a head loss due to friction of ten feet using equation (3), I'm not surprised if the measured loss is seven feet or 15 feet. This sort of imprecision is an aspect of the real-world engineering aspect.

Referring to equation (3) above, note that the head loss is proportional to the pipe diameter to the 5th power. But why?

- Pipe cross-sectional is proportional to the diameter, squared.
- Pipe velocity is proportional to the pipe cross-sectional area.
- Head loss is proportional to the flowing velocity squared. (Equation 3).
- Head loss is also proportional to the ratio of the pipe's cross-sectional area, divided by the pipe wetted area. Meaning smaller inside diameter pipes, have more frictional drag, at a given velocity, than does a larger inside diameter pipe.

Thus, if a four-inch pipe, flowing at a rate of 50 GPM, with a head loss of 10 feet per mile of line, is changed to a two-inch pipe, also flowing at 50 GPM, the head loss would then increase to:

- (10 feet) • $(4 \div 2)^5$ = 320 feet!

For orientation, process piping in refineries and petro-chemical plants, is typically designed for a head loss of about two or three feet, per 100 equivalent feet, or about 100 feet of head loss per mile of line.

Note that the above calculations do not include changes in elevation of the piping, and especially do not include what I call acceleration losses (i.e., converting head to velocity). Only too often I've found that much of the head loss in 1,000 feet of six-inch refinery process piping, connecting two process units, was due to a single foot of three-inch pipe, in a remote and forgotten section of the pipe rack. When I complained to the piping engineer about the three-inch pipe, he said, "But Norm, it's only one foot long out of a 1,000 feet of piping."

FACTOR AFFECTING ORIFICE COEFFICIENTS

The 0.18 orifice coefficient shown in equation (1), is a theoretical coefficient that includes no losses for friction or turbulence. A real-world coefficient is between 0.3 to 0.6. If the opening is smooth, contoured, or rounded, the 0.3 value is appropriate. If the opening is a sharp-edged orifice, the 0.6 value would be corrected. A hole in a steel plate punched down, might have half the pressure drop, as a hole that has been punched down.

Strange to say, a hole in a thick steel plate (½") would have perhaps 60% of the pressure drop, as a hole in a thin steel plate (1/10"), for the same flow of fluid through the hole (i.e., constant hole velocity).

COMPRESSIBLE FLUIDS

The ideas expressed above apply to vapors like air, steam, and natural gas. Except there's an added complication, because vapors or gasses are compressible. When such vapors pass through a hole at an increased velocity (i.e., they accelerate), some of the energy to increase the velocity comes from the heat content of the vapor itself. This is what the subject of Thermodynamics, discussed in a later chapter in this text, is all about.

We calculate the pressure drop for a vapor also using equation (1). However, if we wish to express the pressure drop in terms of inches of water, rather than in terms of inches of the vapor head pressure, we would multiply the calculated pressure drop for the vapor stream by the ratio of:

- (Density of the Vapor) ÷ (Density of the Liquid)

For example, the density of hot air is 0.063 pounds per cubic foot. The density of cold water is 62.3 pounds per cubic foot. Therefore, I would multiply a calculated delta P, for hot air blowing through a hole by:

- 0.063 ÷ 62.3 = 0.001

to obtain the air pressure drop, expressed in inches of water pressure loss.

CHAPTER NINE

AIR COOLERS

I worked for American Oil in Whiting, Indiana, in the 1960s. In those days, we used Lake Michigan as a source of cooling water. Of course, we had exchanger tube leaks that allowed hydrocarbons like benzene and xylene to leak into the lake. But as we used to say, "The solution for pollution is dilution."

In recent times, it's more cost effective to use air coolers. Most of the newer facilities I work on rely on forced draft air coolers. Forced draft means that the fan is located underneath the tube bundle. Induced draft is when the fan is located above the tube bundle, and cooling air is drawn up through the bundle. Induced draft has certain process advantages over forced draft. Most air coolers, however, are forced draft, so that the fan, motor, and drive belt are readily accessible for maintenance.

MEASURING AIR FLOW

I often feel isolated from the rest of mankind. Apparently I'm the only human who has ever observed that air coolers rarely – if ever – develop their rated air flow. The rated air flow, in pounds per hour (lbs/hr) is listed on the vendor's air cooler data sheet.

My method of measuring the air flow is:

- Step 1 – Calculate the process side heat duty in BTU per hour. Let's say I'm condensing 10,000 lbs/hr of butane from 200°F, down to 140°F. The latent heat of condensation of butane is 130 BTU/lb. The sensible heat of butane vapor is 0.60 BTU/LB/°F. Then the process duty is:

- 10,000 • (200°F - 140°F) • (0.60) = 360,000 BTU/HR   (sensible heat)

  +

- 10,000 • (130)   = 1,300,000 BTU/HR   (latent heat)

TOTAL = 1,660,000 BTU/HR

- Step 2 – Measure the average temperature rise of the air. To do this accurately, you will need a long stick and dial thermometer. Tape the thermometer to the stick and read the air outlet temperature at six evenly spaced locations across the top of each bundle. Average the six temperatures. Let's say that average is 130°F and the ambient temperature is 90°F, so that the average temperature rise is 40°F.
- Step 3 – The specific heat of air is 0.25 BTU/LB/°F. So each pound of air can absorb:
    - (40°F) • (0.25) = 10 BTU per pound.
  Since the process side of heat duty was 1,660,000 BTU/HR, the observed air flow is:
    - (1,660,000) ÷ 10 = 166,000 LBS/HR of air.
  I check the vendor's data sheet and note that the design air flow is 408,000 LBS/HR! I've only 40% of the air flow I should have!

AIR RECIRCULATION

I noted before that induced draft fans that pull air through the tube bundle are somewhat better than forced draft. For example, there is no air recirculation in induced draft coolers. I have stood underneath several thousand forced draft air coolers. Without exception, each air cooler has air blowing downwards around its peripheral area. Without exception, each air cooler has air drawn upwards through its central area. Typically, the underside of the screen (underneath the fan blades needed for personal protection) has about 30% of its area, or more, exposed to this reverse air flow.

This recirculated air flows from the discharge of the forced draft fan. Some of the air discharged from the fan is cooling air, blowing up through the process tube bundle. And some of the air discharged from the fan is recirculated air, blowing out of the peripheral area of the screen. This recirculated air does not contribute to cooling.

As all of the thousands of forced air coolers I've seen suffer from air recirculation, I've concluded that it is an intrinsic characteristic of forced draft air coolers. Yet I have never seen any allowance made for air recirculation on any of the thousands of air cooler vendor equipment data sheets I have used as a process design engineer.

Is this an oversight by the equipment vendor? Apparently so. It seems that the sizing of the forced draft fan is based on:

- The air pressure drop through the tube bundle, assuming the exterior of the tubes are clean.
- The required air flow based on the process side duty.
- The specified air inlet and outlet temperatures.

The portion of the air flow from the discharge of the fan that is inevitably recirculated is not included in the total design air flow. Other factors that reduce air flow are:

- Fouling on the exterior fins of the tubes.
- Belts slipping.
- Fan blade pitch not at the optimum (i.e., 20°–25°).
- Larger than design vane tip to shroud clearance.
- Leaks around exterior of the tube bundle.

I'll discuss these potential problems later. But for now, let's consider the consequences of factors that restrict the flow of air through the tube bundle. These factors are:

- Dirt, bugs, and dust accumulation between the fins on the bottom two rows of tubes.
- Bent fins on the bottom row of fins.

Note that bent fins of the upper row of fins do not restrict air flow. Smashed fins on the top of the upper row of tubes look really bad, but have no effect on air flow. The effect of dirt and bent fins on the bottom rows will not increase air flow DP very much. The total air flow rate remains constant. It's just that the air recirculation rate increases, and the cooling air flow decreases.

The easy way to prove this to yourself is to check amps on your AC fixed speed motor. As the tube bundle fouls, the motor amps will remain the same. Or, you can measure the inlet pressure, in inches of water, to the tube bundle with a bottle of water and six feet of clear plastic tubing.

But even when I've had an air cooler bundle cleaned (as described below), I still experience a large percentage of the air blowing backwards through the screen. I still only calculate 70% - 80% of the design air flow. And this is so, even when the vane tip-to-shroud clearance is a uniform ¼" to ½".

VANE TIP CLEARANCE PROBLEMS

The tip of the fan blade is supposed to be ¼" to ½" away from the I.D. of the shroud. You can see the clearance in a bright light. The path of the fan blade will look like a grayish blur. It can't be seen, however, in dim light.

Often, this clearance is not at all uniform. At 12:00 and 6:00 o'clock orientation, the clearance may be fine. But at 3:00 and 9:00, it may be 2" – 4". This is caused by the fan not being installed absolutely level with the shroud structure, or the shroud becoming elliptical with time.

Increased fan blade tip-to-shroud clearance will encourage air recirculation. You can install a vane-to-tip seal mesh easily and cheaply, to correct this problem. I've done this once in Aruba on a hydrogen plant and gained some air flow.

CLEANING TUBE BUNDLE

I have increased air flow (by measurement) by up to 25%, by water washing the underside of the air cooler. Here's my detailed procedure which I've done three times in the last half century:

- Electrically lock-out motor.
- Tie off fan blade with a strong rope.
- Drop off screen.
- Spray a detergent on the underside of the tubes.
- Wait a few hours for the detergent to soak in.
- Obtain a source of clean water with a pressure of 30 to 50 psig at the elevation of the tube bundle.
- Attach a 12" section of ½" s.s. tubing to the end of the hose.
- Run the water carefully between the bottom rows of tubes to wash off deposits.
- Be careful not to bend the fins with too much water pressure.

You may expect to take an hour or so for each bundle and get hot, wet, and dirty. This is the right way to do the job. But not the Norm Lieberman method, which is:

- Reverse the polarity of the 230-volt motor.
- Turn the fan on.
- The dirt will, more or less, blow off the fins.

Where the dirt will blow off to is problematical.

A less messy method to dislodge dirt deposits is to remove the screen and sweep off the bugs and dust. Water washing from the top down is relatively ineffective, unless an extremely large volume of water is used.

EFFECT OF FOULING ON REVERSE AIR FLOW

A small piece of paper will always stick to the center of the screen, below the forced draft fan. The paper will always blow off the peripheral area of the screen. Finned tubed fouling will increase the reverse or recirculated air flow rate. Thus, the area where the paper sticks to the screen will shrink as fouling progresses. This is a simple tool to monitor the rate of finned tube bundle fouling.

Incidentally, I do not believe that tube bundle fouling occurs to any extent above the first and second rows of tubes. It seems that almost all of the moths and road dust is filtered out by the fins on the bottom two rows. Therefore, there is no need to clean the interior fins of the tube bundle.

EFFECT OF ADDING ROWS OF TUBES

I was working for a refinery in Kansas. They had a problem with lack of condenser capacity in their crude distillation tower overhead. To increase capacity, they purchased a new tube bundle that had six, rather than the four rows in the existing tube bundle. The fan was not altered, as it was deemed too expensive to replace the fan motors.

One might think that the air DP went up by 50%. But that would be wrong. The air DP remained constant. But the air recirculation flow increased dramatically, and the cooling air flow dropped off by about 20%. I recall that in the dead of winter, with ambient temperatures of about 25°F, the average air outlet temperature from the cooler was about 120°F. The incremental condenser duty, as a result of the new, and very expensive, new tube bundle, was too small to measure. Electric supply limitations prevented installation of a larger fan motor and also prevented any increase in fan speed.

CHANGING FAN BLADE TIPS & SPEED

The easy way to get more flow from a fan is to increase the fan blade angle, which is adjustable. It's rather difficult in practice to adjust the blade angle closer than plus or minus 5%. The optimum I've read is 22½°. I've only increased the blade pitch on fans twice from around 10%–15° to 20°–25°. For my trouble, I achieved an increased air flow of maybe 5%. Motor amps also increased by about 10%–15%.

Incidentally, motor amps also increase between summer and winter by roughly 5%, for each decrease in ambient temperature of about 20°F.

Increasing fan blade speed is the really effective way to get more air flow. One problem is the extra allowable torque on the blade. Check with the vendor to see if higher fan speed is permitted. More critically, the fan motor amperage loading will increase with the larger wheel on the motor shaft with the diameter cubed. So increasing the motor wheel diameter by 10% will increase the motor amps by 34%. So check the full limit amp load (i.e., the amp overload trip value), before increasing the wheel or pulley size on the motor.

The other way to increase the air flow is to change the older conventional fan blades for newer, fiber blades with tapered tips. Supposedly, this will increase air flow from 25% to up to 40%. So I tried this on one installation and only increased the cooling air flow by around 10%! What went wrong?

Actually nothing. The increase of 10% of air flow also increased the DP through the finned tubes by 21%:

- Pressure Drop is proportional to velocity squared.

Likely, if I could have avoided a higher air blower discharge pressure, the new fan blades, would have increased the cooling air flow a lot more than just 10%. But, the unavoidable increase in the fan discharge pressure, likely caused most of the incremental air to recirculate, rather than flow through the tube bundle. Also, there was a slight increase in the fan motor amperage load.

SLIPPING BELTS

I was working for a client last summer who was unable to meet their paving asphalt viscosity specification due to poor vacuum in their vacuum tower. The cause of the poor vacuum was excessive vapor load to the first stage ejector, due to poor cooling in the vacuum tower top pumparound air cooler. Almost 12,000 BSD of asphalt product was downgraded to Delayed Coker Feed, at a cost of $25/bbl. That's $300,000 per day!

I had to delay my visit to their plant from Friday to Monday, as I was running Saturday in a 10K race in New Orleans (which I did not win in the 70+ age bracket). When I arrived on-site, I found both belts so loose on the dual cooler fans, that one fan was revolving at about 10% of its design speed and the average temperature rise of the air was 100°F. With ambient air at 95°F that day, it was not too hard to troubleshoot the problem.

The shocking thing about this story is that my client had this problem for years, and it had been getting gradually worse, but we corrected it that same afternoon.

My participation in the 10K race cost my client a million dollars. Had I won, my prize would have been a large plastic glass. Incidentally, tightening the loose belts doubled the fan motor amp loads, but they were still significantly below the FLA (full limit amp) point for the motors.

An additional, but relatively minor problem with the bundle was that air was by-passing around the edges of the bundle, where it was not properly aligned with its support. This was fixed with sections of aluminum sheeting.

AIR HUMIDIFICATION

Continuous wetting of an air cooler will degrade, corrode, and foul the air cooler aluminum fins. However, in areas where the relative humidity is moderate (not in New Orleans) misting small amounts of water below the forced draft fans will cool off the air by an appreciable amount. I've measured the effect of air humidification in Lithuania and New Jersey. In both locations, the reduction of the process outlet temperature was 6°F to 7°F. A typical air cooler, using 500,000 lbs/hr of air, might require three or four GPM of water distributed through six to eight mist nozzles purchased at your local hardware store. It's best to use clean boiler feed water or steam condensate. The temperature of the water is irrelevant. Only use the mist during the heat of the day. If the fins are actually wetted by the mist, you are using too much water pressure or too many mist nozzles.

INDUCED DRAFT FANS

These fans are located on top of the tube bundle. Therefore, there is no possibility of reverse air or air recirculation. Also, cleaning the underside of the tube bundle can be done very easily, without even shutting off the fans.

The only potential problems I can envision with induced draft fans are air leaks between the top of the tube bundle and the fan inlet and the air being drawn around the fan blade tips and the I.D. of the shroud. As in forced draft fans, vane tip seals can be used to minimize these latter problems.

CHAPTER TEN

EXTRACTING WORK FROM STEAM

In 1968 I was working for the long since forgotten American Oil Company in Texas City on their Delayed Coker. To cut the coke out of the giant coke drums, a high-pressure water drill stem was slowly lifted by means of an air operated motor. The source of the air to turn the motor was the plant air system. Plant air was about 60 psig and ambient temperature. The plant air, having done its job of spinning the turning motor, was exhausted to the atmosphere. The thing I noticed about the air lift motor was that it was iced over. Now this was Texas City in July, not the Arctic in December. It was 90% humidity and 95°F. Why the ice?

Did you know that the fundamental nature of the universe is hidden from us? For example, the energy that drives an air motor does not come from the pressure of the air. It can't! Because low-pressure air contains more energy than high-pressure air, if the air is at the same temperature at both pressures, and if the air is not moving! Meaning, there's no kinetic energy stored up in the air.

For the molecules of air to move further away from each other takes energy – potential energy. Does that mean that if we allow air to expand from a little box at 60 psig into a big box at two psig, the larger box will be colder? It certainly will. Maybe that partly explains why the air motor in Texas City got cold. But it certainly does not explain where the energy came to spin the motor.

The energy to spin the lift motor came not from the pressure of the air, but from the heat content of the air. Meaning, the source of energy to lift the coke drum drill stem came from the temperature of the air!

AN ISENTROPIC EXPANSION

Forget about the word "Isentropic." Go out to your car. Depress the tire air valve. Feel the cool air blowing past your fingers? Most of that cooling is due to the conversion of temperature into velocity. Or the conversion of **"Enthalpy into Kinetic Energy."** This is called by the evil professors, who wrote horrible books about Thermodynamics, an Isentropic Expansion.

What source of energy in the air caused the air motor in Texas City to spin? Not the pressure of the air, but the temperature of the air! So it works like this:

- Step 1 – The 60 psig air expands to atmospheric pressure though a nozzle inside the air motor.
- Step 2 – As the air expands, its velocity increases to thousands of feet per second.
- Step 3 – The speeding air, which now contains a lot of kinetic energy or momentum (mass times velocity) strikes the spinning rotor of the air motor.
- Step 4 – The momentum of the air is transferred to the motor's rotor, in the form of rotational energy or centrifugal force.
- Step 5 – The air slows down as it loses momentum, and is vented back to the atmosphere.

But what would happen if the rotor jams and the air can't transfer any of its kinetic energy or its momentum to the rotor? Will the air heat up, back to its starting, ambient temperature of 90°F? Yes it will! Exactly so!

The exterior ice I saw on the outside of the air-operated motor was an indication of the enthalpy, or the heat content, or the temperature of the air, being efficiently converted into work to spin the motor and lift the coke drum drill stem.

STEAM TURBINES

More energy is consumed in steam turbines than all the cars in the world! Why?

Because steam turbines are used to make electricity, most of which still comes from coal-fired and nuclear power plants. For me, because of an accident at birth (the nurse dropped a clipboard on my head), I spend much of my time troubleshooting steam turbine drivers in refineries and petrochemical plants in Texas and Louisiana. When I tell the operators that the energy to drive the steam turbine comes not from the pressure of the steam, but from its heat, they don't believe me. So, using my infrared temperature gun, I'll check the motive steam supply temperature. Typically 400°F. Then, I check the temperature on the turbine case. It's 200°F.

"See, guys," I'll say, "It's the temperature of the steam, converted to velocity that spins the turbine wheel. Not its pressure."

And the operators say, "Well, you damn Yankee, why don't you go back to where you belong, to New York City?"

And I say, "Hey! I come from the South too! South Brooklyn!"

THE POTENTIAL ENERGY OF STEAM

Rick Jones, the tech service manager of Ineos Chemicals in Houston, has questioned me on my idea that the energy that steam uses to drive a turbine comes from its heat content, and not from its pressure.

"Rick," I explained, "it's true that when I have a hydroelectric power station, I use the water level in the dam to generate electricity. Let me show you how to calculate the amount of electricity I can get from water. First, note, Rick, that:

- 770 ft-lbs = one BTU
- 3,000 BTU/HR = one kilowatt
- One psi = 2.31 ft. of water

Let's say I've got one pound per hour of water, that's flowing over a dam of 1,000 feet high. So, that amount of electricity I could make from the pound of water is:

- $$\frac{(1,000 \text{ ft}) \cdot (\text{one pound/hr})}{(770) \cdot (3,000)} = 0.00044 \text{ KW}$$

"Norm, you're being stupid," Rick observed. "You don't build a dam a 1,000 feet high for a single pound of water.

"But let's say, Rick, I take a pound of steam at 420 psig and 500°F, and let it down through a steam nozzle to atmospheric pressure then…"

"Hey, Norm, is the 420 psig steam pressure, is that like 1,000 ft. of head for water?"

"Yeah, Rick."

- (420 psi) • (2.31 ft/psi) = 1,000 ft.

"So, Rick, when I de-pressure the steam through a nozzle, that's called an Isentropic Expansion. And if you look on your **Mollier Diagram** in the back of your Steam Tables, you'll see that I can extract about 210 BTU/LB of heat, which can be converted into work or power:

- (210 BTU/LB) • (1.0 LB/HR) ÷ (3,000 BTU/KW) = 0.070 KW

"Wow, Norm. That's 160 times as much electric power that I can get from the heat content or enthalpy of the steam, than I could extract from its pressure alone!"

"Yeah, Rick! You need a whole river of water to generate a few megawatts of electricity from a hydropower station, but only a few hundred GPM of boiler feed water for a coal-fired steam power station."

"Steam is great. I've always been a big fan of steam," said Rick.

"But, Rick, you realize that all of my calculations assume 100% efficiency. In reality, you should use maybe 70% to 80%, to allow for losses."

"Gee, how come they didn't teach us about this stuff at Purdue University? It's interesting!"

"They did, Rick. It was called **Thermodynamics**. I guess you were too interested in trying to meet girls at Purdue to waste time on thermo?"

CONDENSING STEAM AT LOW PRESSURE

In the above example, the steam was expanded through a nozzle from a high pressure, down to atmospheric pressure of 14.7 psia. But suppose I was to expand the steam into a chamber which operated under a vacuum. Say at a pressure of 0.10 atmospheres or 1.47 psia. Well, since the volume of a vapor is inversely proportional to pressure:

- $V \sim 1/P$

Then, it follows that the volume of the steam would increase by ten times. The velocity of the steam would also increase by ten times. And, the amount of work we could extract from each pound of steam would also increase a lot. Not by ten times, but by around 50%. You can see that on your Mollier Diagram. Drop straight down the lines of constant entropy until you intersect the constant pressure line of 1.5 psia.

Incidentally, to convert from psia to inches of Mercury (Hg) absolute, multiply by two. To convert to mm of Hg, multiply psia by 52. I always just use mm of Hg (absolute) in my work, because atmospheric pressure varies so much on different parts of the earth's surface and with barometric pressure.

THE MEANING OF ENTROPY

For you and me, as engineers and operating technicians, entropy only relates to how we expand high-pressure steam into low-pressure steam. When I have done everything possible to convert the heat in steam (or enthalpy), into kinetic energy (or velocity), this is called an Isentropic Expansion. To you and me then, an Isentropic expansion means I have expanded the steam, or let down its pressure, so that I have preserved the ability of the steam to do **Work**. And that ability to do work comes not from the pressure of the steam – but from the heat of the steam.

Does this mean that if I slow the steam back down, but without extracting any work from it, it will heat back up? And the steam's pressure will also go back up? Yes! As long as there are no frictional losses, the expansion of the steam is fully reversible.

The person who first used the idea that you could extract a lot of work from the heat content of steam was James Watt, who did not invent the steam engine. Mr. Watt invented the external steam condenser, which allowed the steam engine exhaust to be condensed under vacuum. James Watt is my hero. He came from a poor family, and using technology, made lots and lots of money.

He sprayed cold water into the engine's steam exhaust, to create a partial vacuum, and called the spray chamber the barometric condenser. I use that same idea in my work all the time to directly de-superheat vapors with a cold spray of liquid to promote condensation. I have the same objective as James Watt. That is, to make lots of money.

The Norm Lieberman Refinery Troubleshooting Seminar Video Presentation

In 1988 I presented at the Texaco Refinery in Anacortes, Washington State, a three-day seminar for a group of senior plant operators and engineers. As refinery technology has not altered in the last 30 years, the material in these videos still is the same as I present currently in my seminars. The main change is that I have become older.

This material is not suitable for new operator training. The Texaco participants in the seminar were mainly experienced plant operators and engineers.

Norm Lieberman Refinery Troubleshooting Seminar

A three-day Troubleshooting Seminar presented for Process Operators and engineers, covering basic concepts of unit operations.

The set comprises 23 lessons:

Section I – Distillation	10 DVDs	$3,900.00
Section II – Heat Exchange, Hydraulics & Vacuum Systems	13 DVDs	$3,900.00
Complete Set – Sections I & II	23 DVDs	$6,800.00

Four copies of the seminar workbook are provided with each set. The workbook contains all copies of drawings used in the seminar.

Further reference text for the seminar is *A Working Guide to Process Equipment*, 4th Edition, McGraw-Hill.

Please send all inquiries and orders to:

PROCESS CHEMICALS, INC.
5000 West Esplanade Ave.
PMB 267
Metairie, Louisiana 70006, USA
Ph: (504) 887-7714
Fax: (504) 456-1835
norm@lieberman-eng.com

A Norm Lieberman Refinery Troubleshooting Seminar

INTRODUCTION   10 min

Section I – Distillation

DVD 1	Fundamentals of Distillation	50 min
DVD 2	Tray Deck Flooding: Part I	58 min
DVD 3	Tray Deck Flooding: Part II	30 min
DVD 4	Variables Affecting Distillation Tower Performance	21 min
DVD 5	Optimizing Fractionator Operating Pressure	40 min
DVD 6	Troubleshooting Vacuum, Crude, & Coker Fractionators	58 min
DVD 7	Side Stream Draw Rates & Quality	45 min
DVD 8	Side Stream Steam Stripping	32 min
DVD 9	Level Control	47 min
DVD 10	Distillation Tower Troubleshooting Examples	58 min

Section II – Heat Exchange, Hydraulics, & Vacuum Systems

DVD 11	Heat Exchanger Internals	57 min
DVD 12	Reboiler Hydraulic Problems	37 min
DVD 13	Fired Heaters	53 min
DVD 14	Air Preheaters, Draft Related Problems & Tube Side Coking	49 min
DVD 15	Partial Condensers & Air Coolers	40 min
DVD 16	Condensation & Distillation Overhead Systems	51 min
DVD 17	Condenser Subcooling & Steam Turbine Operation	32 min
DVD 18	Exchanger Back Flushing & Cooling Tower Operation	60 min
DVD 19	Vacuum Tower & Steam Ejector Problems	60 min
DVD 20	Surface Condensers	56 min
DVD 21	Steam Turbine Air Blower & Centrifugal Pumps	52 min
DVD 22	Centrifugal Compressors	43 min
DVD 23	Reciprocating Compressors	39 min

Index

Also of Interest

Check out these other related titles from Scrivener Publishing

From the Same Author

Troubleshooting Vacuum Systems: Steam Turbine Surface Condensersand Refinery Vacuum Towers, by Norman P. Lieberman. ISBN 978-1-118-29034-7. Vacuum towers and condensing steam turbines require effective vacuum systems for efficient operation. Norm Lieberman's text describes in easy to understand language, without reference to complex mathematics, how vacuum systems work, what can go wrong, and how to make field observations to pinpoint the particular malfunction causing poor vacuum. *NOW AVAILABLE!*

Other Related Titles from Scrivener Publishing

An Introduction to Petroleum Technology, Economics, and Politics, by James Speight, ISBN 9781118012994. The perfect primer for anyone wishing to learn about the petroleum industry, for the layperson or the engineer. *NOW AVAILABLE!*

Ethics in Engineering, by James Speight and Russell Foote, ISBN 9780470626023. Covers the most thought-provoking ethical questions in engineering. *NOW AVAILABLE!*

Fundamentals of the Petrophysics of Oil and Gas Reservoirs, by Buryakovsky, Chilingar, Rieke, and Shin. ISBN 9781118344477. The most comprehensive book ever written on the basics of petrophysics for oil and gas reservoirs. *NOW AVAILABLE!*

Mechanics of Fluid Flow, by Basniev, Dmitriev, and Chilingar. Coming in September 2012, ISBN 9781118385067. The mechanics of fluid flow is one of the most important fundamental engineering disciplines explaining both natural phenomena and human-induced processes. A group of some of the best-known petroleum engineers in the world give a thorough understanding of this important discipline, central to the operations of the oil and gas industry.

Zero-Waste Engineering, by Rafiqul Islam, ISBN 9780470626047. In this controversial new volume, the author explores the question of zero-waste engineering and how it can be done, efficiently and profitably. *NOW AVAILABLE!*

Sustainable Energy Pricing, by Gary Zatzman, ISBN 9780470901632. In this controversial new volume, the author explores a new science of energy pricing and how it can be done in a way that is sustainable for the world's economy and environment. *NOW AVAILABLE!*

Emergency Response Management for Offshore Oil Spills, by Nicholas P. Cheremisinoff, PhD, and Anton Davletshin, ISBN 9780470927120. The first book to examine the Deepwater Horizon disaster and offer processes for safety and environmental protection. *NOW AVAILABLE!*